The Beltrami Equation

Memoirs
of the
American Mathematical Society

Number 893

The Beltrami Equation

Tadeusz Iwaniec
Gaven Martin

January 2008 • Volume 191 • Number 893 (third of 5 numbers) • ISSN 0065-9266

American Mathematical Society
Providence, Rhode Island

2000 *Mathematics Subject Classification.* Primary 35J15, 35J70.

Library of Congress Cataloging-in-Publication Data

Iwaniec, Tadeusz.
  The Beltrami equation / Tadeusz Iwaniec, Gaven Martin.
    p. cm. — (Memoirs of the American Mathematical Society, ISSN 0065-9266 ; no. 893)
  "Volume 191, number 893 (third of 5 numbers)."
  ISBN 978-0-8218-4045-0 (alk. paper)
  1. Conformal mapping. 2. Differential equations, Partial. 3. Geometry, Non-Euclidean. I. Martin, Gaven. II. Title.
  QA360.I935   2007
  515'.3533—dc22   2007060558

Memoirs of the American Mathematical Society

This journal is devoted entirely to research in pure and applied mathematics.

Subscription information. The 2008 subscription begins with volume 191 and consists of six mailings, each containing one or more numbers. Subscription prices for 2008 are US$675 list, US$540 institutional member. A late charge of 10% of the subscription price will be imposed on orders received from nonmembers after January 1 of the subscription year. Subscribers outside the United States and India must pay a postage surcharge of US$38; subscribers in India must pay a postage surcharge of US$43. Expedited delivery to destinations in North America US$53; elsewhere US$130. Each number may be ordered separately; *please specify number* when ordering an individual number. For prices and titles of recently released numbers, see the New Publications sections of the *Notices of the American Mathematical Society*.

Back number information. For back issues see the *AMS Catalog of Publications*.

Subscriptions and orders should be addressed to the American Mathematical Society, P. O. Box 845904, Boston, MA 02284-5904, USA. *All orders must be accompanied by payment.* Other correspondence should be addressed to 201 Charles Street, Providence, RI 02904-2294, USA.

Copying and reprinting. Individual readers of this publication, and nonprofit libraries acting for them, are permitted to make fair use of the material, such as to copy a chapter for use in teaching or research. Permission is granted to quote brief passages from this publication in reviews, provided the customary acknowledgment of the source is given.

Republication, systematic copying, or multiple reproduction of any material in this publication is permitted only under license from the American Mathematical Society. Requests for such permission should be addressed to the Acquisitions Department, American Mathematical Society, 201 Charles Street, Providence, Rhode Island 02904-2294, USA. Requests can also be made by e-mail to reprint-permission@ams.org.

Memoirs of the American Mathematical Society is published bimonthly (each volume consisting usually of more than one number) by the American Mathematical Society at 201 Charles Street, Providence, RI 02904-2294, USA. Periodicals postage paid at Providence, RI. Postmaster: Send address changes to Memoirs, American Mathematical Society, 201 Charles Street, Providence, RI 02904-2294, USA.

© 2008 by the American Mathematical Society. All rights reserved.
Copyright of individual articles may revert to the public domain 28 years after publication. Contact the AMS for copyright status of individual articles.
This publication is indexed in *Science Citation Index*®, *SciSearch*®, *Research Alert*®, *CompuMath Citation Index*®, *Current Contents*®/*Physical, Chemical & Earth Sciences*.
Printed in the United States of America.
∞ The paper used in this book is acid-free and falls within the guidelines established to ensure permanence and durability.
Visit the AMS home page at http://www.ams.org/

10 9 8 7 6 5 4 3 2 1   13 12 11 10 09 08

In memory of Eugenio Beltrami (1835-1900)

Contents

Chapter 1. Introduction	1
Chapter 2. Quasiconformal Mappings	5
2.1. Analytic Definition of Quasiconformality	5
2.2. The Beltrami Equation	6
2.3. Radial Stretchings	7
2.4. Classical Regularity Theory	8
Chapter 3. Partial Differential Equations	11
3.1. The Transformation Formula	11
3.2. A Fundamental Example	12
3.3. The Construction	13
3.4. Cavitation and Riemann Surfaces	15
Chapter 4. Mappings of Finite Distortion	17
4.1. Orlicz–Sobolev Spaces	18
4.2. Monotonicity	21
4.3. A Class of Orlicz Functions	22
4.4. The Monotonicity Theorem	23
4.5. Modulus of Continuity	24
Chapter 5. Hardy Spaces and BMO	27
5.1. Mollifiers	27
5.2. Hardy-Orlicz Spaces	28
5.3. BMO	29
5.4. $L \log L$–Integrability	30
5.5. Liouville Type Theorems	30
Chapter 6. The Principal Solution	33
6.1. Solutions	33
6.2. Uniqueness of Principal Solutions	34
6.3. Stoilow Factorization	35
Chapter 7. Solutions for Integrable Distortion	39
7.1. Distortion in the Exponential Class	41
7.2. An Example	42
7.3. Results	43
7.4. Distortion in the Subexponential Class	45
7.5. An Example	45
7.6. Further Generalities	47
7.7. Existence Theory	48

7.8.	Global Solutions	60
7.9.	Holomorphic Dependence	64
7.10.	Examples and Non-Uniqueness	67
7.11.	Equations in the Plane	73
7.12.	Compactness	77
7.13.	Removable Singularities	79
7.14.	Final Comments	80
Chapter 8.	Some Technical Results	81
8.1.	The Divergence Condition	81
8.2.	Integration by Parts	84
8.3.	Higher Integrability	86
Bibliography		89

Abstract

The "measurable Riemann Mapping Theorem" (or the existence theorem for quasiconformal mappings) has found a central rôle in a diverse variety areas such as holomorphic dynamics, Teichmüller theory, low dimensional topology and geometry, and the planar theory of PDEs. Anticipating the needs of future researchers we give an account of the "state of the art" as it pertains to this theorem, that is to the existence and uniqueness theory of the planar Beltrami equation, and various properties of the solutions to this equation. The classical theory concerns itself with the uniformly elliptic case (quasiconformal mappings). Here we develop the theory in the more general framework of mappings of finite distortion and the associated degenerate elliptic equations.

We recount aspects of this classical theory for the uninitiated, and then develop the more general theory. Much of this is either new at the time of writing, or provides a new approach and new insights into the theory. Indeed, it is the substantial recent advances in non-linear harmonic analysis, Sobolev theory and geometric function theory that motivated our approach here. The concept of a *principal solution* and its fundamental role in understanding the natural domain of definition of a given Beltrami operator is emphasized in our investigations. We believe our results shed considerable new light on the theory of planar quasiconformal mappings and have the potential for wide applications, some of which we discuss.

Received by the editor Received by the editor July 27, 2000, and in revised form September 10, 2004.

2000 *Mathematics Subject Classification.* Primary 30C60, 35J15, 35J70.

Key words and phrases. Beltrami Equation, quasiconformal, finite distortion, partial differential equations, degenerate elliptic.

The first author was supported in part by grants from the US National Science Foundation.

The second author was supported in part by the NZ Marsden Fund, the Royal Society of NZ (James Cook Fellow) and the NSF.

CHAPTER 1

Introduction

The interplay between Partial Differential Equations (PDEs) and the theory of mappings has a long and distinguished history. Gauss' practical work of geodesic surveying stimulated him to develop the theory of conformal transformations, for mapping figures from one surface to another. For conformal transformation from plane to plane, he used a pair of equations, apparently derived by d'Alembert who first related the derivatives of the real and imaginary part of a complex function in 1746 in his work on hydrodynamics [15] pg 497. These equations have become known as the Cauchy-Riemann equations. Gauss developed the differential geometry of surfaces (1827) emphasizing the intrinsic geometry, with Gaussian curvature defined by measurements within the surface. Gauss also considered geodesic curves within surfaces.

In 1829, Lobachevsky constructed a surface (the horosphere) within his non-Euclidean space, such that the intrinsic geometry within that surface is Euclidean, with geodesic curves being called Euclidean lines. For the converse process, he could only suggest tentatively that, within Euclidean space, the intrinsic geometry of a sphere of imaginary radius was Lobachevskian. But imaginary numbers were then regarded with justifiable suspicion, and he did not propose that as an acceptable model of his geometry within Euclidean space. The most famous work of Beltrami is [16] from 1867. There he showed that Lobachevsky's geometry is the intrinsic geometry of a surface of constant negative curvature, with geodesic curves being called lines in this geometry. Beltrami illustrated various surfaces with constant negative curvature, the simplest of which is the pseudosphere generated by revolving a tractrix around its axis. Beltrami's paper convinced most mathematicians that the geometries of Euclid and of Lobachevsky are logically equivalent. In that work Beltrami used a differential equation corresponding to Gauss's equation. This has come to be known as the Beltrami equation. This major work of Beltrami's appeared 3 years after he took up the geodesy professorship at Pisa after it was turned down by Riemann. Beltrami went on to become one of Italy's foremost mathematicians, primarily as a geometer but making significant contributions in analysis, the theory of heat, and to mechanics.

Of course since those early times, the theory of conformal mappings, analytic functions, Riemann surfaces and so forth has expanded in many different directions, far too numerous to relate, and now lie at the foundation of much of modern mathematics. Moreover practical applications, such as in fluid flow, hydrodynamics and more modern areas of control theory and robotics etc, abound.

In this paper we present the most recent developments concerning certain linear and non-linear equations in the plane, solving Beltrami's equation at the critical point, where the uniform ellipticity bounds are lost. These provide far reaching extensions of Morrey's classical result [**87**] on the existence of homeomorphic L^2 solutions to the Beltrami equation, what has come to be known as the "measurable Riemann mapping theorem" and extend more recent work of David [**33**] and Brakolova–Jenkins [**26**] which we discuss below.

Quasiconformal mappings (the homeomorphic solutions of Beltrami equations) provide a class of mappings which lie "between" homeomorphisms and diffeomorphisms and yet retain many of the features of analytic functions. They were first used in a significant way by Ahlfors in 1935 [**1**] when they proved to be an essential ingredient of his geometric development of Nevanlinna theory via the theory of covering spaces and Riemann surfaces.

Using the Beltrami equation, Bers [**19**] gave a careful exposition of the method of quasiconformal mappings applied to the classical problems of uniformization in complex analysis. Bers gave a proof that every homotopy class contains a locally quasiconformal homeomorphism of one Riemann surface onto another and he also gave complete and independent proofs of the Limit Circle Theorem (uniformization of a closed surface with signature), the Principal Circle Theorem (closed surfaces with symmetries), and the Retrosection Theorem (Schottky uniformization). Additionally he showed how to construct a simultaneous uniformization of two surfaces by means of a quasi-Fuchsian group. Teichmüller theory shows how deformations, or parameterized families, of discrete groups are related to each other via quasiconformal mappings. Implying for instance that the Cantor limit set of a finitely generated Schottky group is quasiconformally equivalent to the usual middle thirds Cantor set. As illustrated in Figure 1. these Cantor sets can be quite complicated.

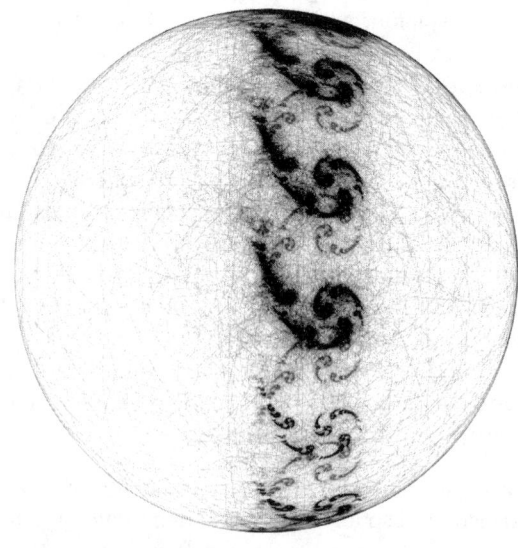

Figure 1. A Cantor limit set of a two-generator Schottky group.

About twenty years ago Sullivan [**104**] introduced the Beltrami equation into the theory of holomorphic dynamics in his solution of the Fatou–Julia problem on wandering domains and revolutionized the field. Since then there has been an explosion of applications in this area, see for instance [**99, 78, 84**] where quasiconformal maps play an essential role in describing families of Julia sets and exhibiting the self similarity of the Mandelbrot set. We also mention the work of Thurston, McMullen and many others [**106, 83**] giving applications of quasiconformal mappings in 3–dimensional topology and geometry.

The more general classes of mappings we shall soon meet, the so called "mappings of finite distortion", are even more flexible than quasiconformal mappings. Among them are the solutions of degenerate Beltrami equations. Many constructions in analysis, geometry and topology rely on limiting processes. The existence, uniqueness and compactness properties of families of mappings with finite distortion make them ideal tools for solving various problems in these areas. For instance in studying deformations of elastic bodies and the related extremals for variational integrals in certain degenerate settings, mappings of finite distortion are often the natural candidates to consider because they are closed under uniform convergence, whereas the limit of diffeomorphisms need not be smooth nor even a homeomorphism. Further applications can be found in our recent joint work [**9**] where a first attempt to study extremal mappings in a more general setting with potential applications to Teichmüller theory. In fact more generally when one studies the boundary of spaces of deformations of conformal dynamical systems one is quickly led to consider mappings which are reasonably regular but are not quasiconformal, such as mappings of finite distortion. Indeed David's results have already found important application in the theory of iteration of rational functions, see for instance Haissinsky and Tan [**46**]. This is illustrated in Figure 2.

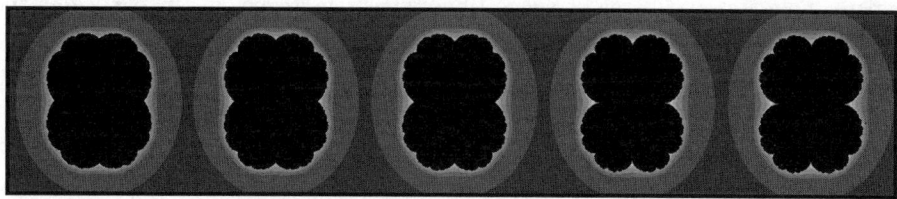

Figure 2. A family of filled in Julia set quasicircles degenerating to a Julia set (the "cauliflower") of finite distortion[1].

The first important results in the theory of mappings of finite, but unbounded, distortion are due to Lehto in 1969 [**72, 73**]. He gave geometric conditions (in terms of moduli of separating annuli) and analytic conditions (involving sophisticated means of distortion functions) to determine sufficient conditions for existence and uniqueness for solutions to the Beltrami equation (defined below at (2.3)) away from a compact singular set where the distortion function may become unbounded. These results are quite difficult to use in practise, though Lehto established a number of the uniform estimates needed to develop the theory. Some years later David (1986)

---

[1]This picture produced using R.L. Devaney's "Mandelbrot Set Explorer", see http://math.bu.edu/DYSYS/explorer/

greatly refined these results and extended their utility by proving existence and uniquess in the case that the exponential of the distortion function is L^p–integrable (he used a slightly different but equivalent condition). Indeed David proved more than he stated and achieved $W^{1,2}$–regularity for sufficiently large p, quite close to the results of our §11. He also established the important modulus of continuity estimates. David's methods were entirely analytic and in spirit close to the methods we employ here. Finally Brakalova and Jenkins (1998) established existence in the case of subexponentially integrable distortion. Their methods were geometric using the modulus of curve families (and so modulus of continuity arguments). Little analytic regularity was obtained. Their existence results can also be improved using our factorisation trick in §§12 & 13.

In this monograph we synthesize and extend these latter two approaches to studying the degerate Beltrami equation employing analytic techniques to discuss extistence, regularity and uniqueness in the degenerate setting obtaining results which are very close to optimal. Our approach uses the substantial advances made in Harmonic analysis in recent times as well as important new developments concerning properties of Jacobians. A discussion of a number of these important, but quite technical, properties of Jacobian determinants has been relegated to the appendices.

While being fully aware of the well–known aphorism of Yogi Berra "making predictions is hard, especially about the future" we hope, and indeed expect, that the developments in the theory of the planar Beltrami equation as presented here will have substantial applications elsewhere. We discuss a couple of possibilities in §17. Some problems and potential future directions are discussed in [**61**].

Finally we would like to thank the anonymous referee who carefully read the paper and made a number of suggestions which improved it. In particular the referee pointed out Lehto's papers, observed the stronger results David obtained and saw how Brakalova and Jenkins results could be improved to what we obtain.

CHAPTER 2

Quasiconformal Mappings

Let Ω and Ω' be planar domains, $f : \Omega \to \Omega'$ a homeomorphism and let $z = x + iy \in \Omega$ and $r < \mathrm{dist}(z, \partial\Omega)$. Quasiconformal mappings are principally mappings of "bounded distortion". If we wish to measure the distortion of f at z it is natural to introduce the quantity

$$(2.1) \qquad H_f(z) = \limsup_{r \to 0} \frac{\max_{|h|=r} |f(z+h) - f(z)|}{\min_{|h|=r} |f(z+h) - f(z)|}$$

This is the *linear distortion function*. If f is conformal, then $H_f(z) = 1$. Indeed the converse is also true. This reflects the fact that infinitesimally conformal mappings preserve angles and "roundness". If f is a diffeomorphism, then it is easy to see that $H_f(z)$ is finite, but not necessarily uniformly bounded as z approaches $\partial\Omega$. A diffeomorphism $f : \Omega \to \Omega'$ has *bounded distortion* if $H_f(z) \leq K$ for all $z \in \Omega$ and some constant $1 \leq K < \infty$.

2.1. Analytic Definition of Quasiconformality

Unfortunately this geometric definition, while aesthetically pleasing, is difficult to work with. These days the following analytic definition of quasiconformality is more common.

Definition. A homeomorphism $f : \Omega \to \Omega'$ is called *K–quasiconformal* if f lies in the Sobolev class $W^{1,2}_{loc}(\Omega, \mathbb{C})$, of complex valued functions whose first order partial derivatives are locally square integrable, and if its directional derivatives satisfy

$$(2.2) \qquad \max_\alpha |\partial_\alpha f(z)| \leq K \min_\alpha |\partial_\alpha f(z)|$$

for almost every $z \in \Omega$.

We point out that there is no gain in generality if we simply assume that $f \in W^{1,1}_{loc}(\Omega, \mathbb{C})$. This is because a homeomorphism of this Sobolev class has locally integrable Jacobian and so the distortion inequality (2.2) implies that $f \in W^{1,2}_{loc}(\Omega, \mathbb{C})$.

It is not difficult to see the relationship between quasiconformality and bounded distortion for diffeomorphisms. It was in this setting that planar quasiconformal mappings were first studied around 1928 by Grötzsch [45]. The term "quasiconformal" was coined by Ahlfors in 1935 [1, 2] when this class of mappings proved an integral tool in his geometric development of Nevanlinna theory, based on the "length–area" method. Teichmüller found a fundamental connection between quasiconformal mappings and quadratic differentials in his studies of extremal mappings between Riemann surfaces [105] around 1939. Developments of the length area method lead to the definition of quasiconformal mappings in terms of the distortion of the modulus of curve families by Pflüger [93] and then were systematically studied in their own right by Ahlfors in 1953 [4].

The class of quasiconformal diffeomorphisms is not closed under uniform limits. Thus the generalization to Sobolev spaces is necessary if one is to solve various extremal problems and make use of limiting processes. After making this generalization we find the limit of a suitably normalized sequence of quasiconformal mappings is either quasiconformal or constant. In this setting the class of quasiconformal mappings becomes more flexible and has a greater range of applications.

The equivalence between the geometric definition and the analytic definition was shown by Gehring and Lehto in 1959 [40]. The connection between quasiconformal mappings, Teichmüller theory and quadratic differentials has been intensively investigated by Ahlfors–Bers, Reich–Strebel, Lehto and others, see [74] and the references therein.

2.2. The Beltrami Equation

There is another route to the theory of planar quasiconformal mappings and it is this route that we will largely focus on in the body of this paper.

Directly from the analytic definition we see that there is a measurable function μ defined in Ω such that

$$(2.3) \qquad \overline{\partial} f(z) = \mu(z)\, \partial f(z)$$

where $\overline{\partial} = \frac{1}{2}(\partial_x + i\partial_y)$ and $\partial = \frac{1}{2}(\partial_x - i\partial_y)$. Indeed

$$(2.4) \qquad \|\mu\|_\infty = \frac{K-1}{K+1} < 1$$

Equation (2.3) is called the *complex Beltrami equation* and K provides the ellipticity bounds for this differential equation. Notice that when $\mu = 0$, or equivalently $K = 1$, we obtain the usual Cauchy–Riemann system. The Beltrami equation, as we mentioned earlier, has a long history. Gauss first studied the equation, with smooth μ, in the 1820's while investigating the problem of existence of isothermal coordinates on a given surface. The complex Beltrami equation was intensively studied by Morrey in the late 1930's, and he established the existence of homeomorphic solutions for measurable μ [87, 88]. Lehto points out, [74] pp. 24, that it took another 20 years before Bers recognized that homeomorphic solutions are quasiconformal mappings in [17].

The function $\mu_f = \partial f / \overline{\partial} f$ is called the *Beltrami coefficient* of f or the *complex dilatation* of f.

Suppose that f is an orientation preserving linear mapping of $\mathbb{C}$. Then a little geometric calculation will show that $f(z) = az + b\overline{z}$, $|a| > |b|$, $\mu_f(z) = b/a$ and f maps the unit circle to an ellipse and the ratio of the major and minor axes of this ellipse is

$$K = \frac{|a|+|b|}{|a|-|b|} = \frac{1+|\mu_f|}{1-|\mu_f|}$$

In this way we may view the Beltrami coefficient of a quasiconformal $f : \Omega \to \mathbb{C}$ as defining a measurable ellipse field on the domain Ω via the affine approximation to f at each point z_0. Indeed the quasiconformal mapping f is differentiable at almost every $z_0 \in \Omega$ and we have

$$f(z) = z_0 + \partial f(z_0)(z-z_0) + \overline{\partial} f(z_0)(\overline{z-z_0}) + o(|z-z_0|)$$

This point of view is especially common in the field of holomorphic dynamics.

Studying quasiconformal mappings via the Beltrami equation is a particularly valuable idea because from this point of view the mapping is the solution of an elliptic equation and as such enjoys various nice properties not obvious from the definition. Before getting too far into the theory here it is important to have a concrete example in hand.

2.3. Radial Stretchings

The first example anyone usually meets in the theory is the radial stretching. Fortunately it is also one of the most useful examples.

Let $\rho = \rho(t)$ be a piecewise differentiable function defined and valued in the interval $[0, +\infty]$. We assume that ρ is either increasing or decreasing. A mapping $h : \mathbb{C} \to \mathbb{C}$ of the form

$$(2.5) \qquad h(z) = \rho(r)e^{i\theta}, \qquad z = re^{i\theta},$$

is called a radial stretching. Note that h preserves (respectively reverses) the orientation if ρ is increasing (decreasing).

We calculate the complex partials of h at points where the derivative $\dot{\rho}(r)$ exists to find

$$(2.6) \qquad h_z = \frac{1}{2}\left[\dot{\rho}(|z|) + \frac{1}{|z|}\rho(|z|)\right]$$

$$(2.7) \qquad h_{\bar{z}} = \frac{1}{2}\left[\dot{\rho}(|z|) - \frac{1}{|z|}\rho(|z|)\right]\frac{z}{\bar{z}}$$

In the orientation preserving case we find that h satisfies the Beltrami equation

$$h_{\bar{z}} = \mu(z) h_z$$

with the Beltrami coefficient μ given by

$$(2.8) \qquad \mu(z) = \frac{|z|\dot{\rho}(|z|) - \rho(|z|)}{|z|\dot{\rho}(|z|) + \rho(|z|)} \frac{z}{\bar{z}}$$

and hence

$$(2.9) \qquad K(z) = \frac{1 + |\mu(z)|}{1 - |\mu(z)|} = \max\left\{\frac{|z|\dot{\rho}(|z|)}{\rho(|z|)}, \frac{\rho(|z|)}{|z|\dot{\rho}(|z|)}\right\}$$

The Jacobian determinant is given by

$$(2.10) \qquad J(z, h) = |h_z|^2 - |h_{\bar{z}}|^2 = \frac{\dot{\rho}(|z|)\rho(|z|)}{|z|}$$

Notice that if h preserves orientation, $J(z, h)$ is always locally integrable. Indeed

$$\int_{|z| \leq \lambda} J(z, h) = 2\pi \int_0^\lambda \dot{\rho}(t)\rho(t) dt$$

$$= \pi \int_0^\lambda d\rho^2 = \pi \rho^2(\lambda)$$

this latter term being the area of the disk of radius $\rho(\lambda)$ of course. Finally, we compute the operator norm of the differential matrix

$$(2.11) \qquad |Dh(z)| = |h_z| + |h_{\bar{z}}| = \max\{\dot{\rho}(|z|), |z|^{-1}\rho(|z|)\}$$

2.4. Classical Regularity Theory

Recall that our definition of a quasiconformal mapping f requires that f lies in the Sobolev space $W_{loc}^{1,2}(\Omega)$. Consider the radial stretching defined on the unit disk **B** using

(2.12) $$\rho(t) = t^\alpha, \qquad 0 < \alpha < 1.$$

We see that

(2.13) $$K(z,h) = \frac{1}{\alpha}, \qquad |Dh(z)| = t^{\alpha-1}, \qquad J(z,h) = \alpha\, t^{2\alpha-2}$$

and hence $h \in W^{1,p}(\mathbf{B})$ for all p such that

$$1 \leq p < \frac{2}{1-\alpha} = \frac{2K}{K-1}$$

and in no better Sobolev space. Notice that

$$\|\mu_f\|_\infty = \frac{1-\alpha}{1+\alpha}$$

For various reasons, this mapping was considered extremal and the above conclusion was conjectured to hold in complete generality. Let us discuss this a little before revealing the complete solution.

In studying solutions to the Beltrami equation an operator, analogous to the Hilbert transform, was introduced. This operator is now known as the Beurling–Ahlfors transform [5]. It is defined as a singular integral of Calderòn–Zygmund type,

(2.14) $$S\omega(z) = \frac{1}{2\pi i} \iint_{\mathbf{C}} \frac{\omega(\zeta)d\zeta \wedge d\bar{\zeta}}{(\zeta - z)^2}$$

for all functions $\omega \in L^p(\mathbb{C})$, $1 < p < \infty$. More precisely, the integral is understood by means of the Cauchy principal value. We refer to Stein's book [102] as a general reference for general facts concerning singular integrals. The operator $S : L^p(\mathbb{C}) \to L^p(\mathbb{C})$ is bounded for all $1 < p < \infty$ and is an isometry in $L^2(\mathbb{C})$. We denote the p-norms of the operator S by $\|S\|_p$, so $\|S\|_2 = 1$. The characteristic property of this operator, and the property which makes it very important to complex analysis, is that it intertwines the ∂ and $\bar{\partial}$ derivatives [1].

(2.15) $$S \circ \frac{\partial}{\partial \bar{z}} = \frac{\partial}{\partial z}$$

It was Bojarski [21, 22] who first gave the elegant analytic proof of the existence and uniqueness for the Beltrami equation using this operator from which he developed the L^p theory of planar quasiconformal mappings.

Closely related to the operator S is the complex Riesz potential

(2.16) $$T\omega(z) = \frac{1}{2\pi i} \iint_{\mathbf{C}} \frac{\omega(\zeta)d\zeta \wedge d\bar{\zeta}}{(\zeta - z)}$$

---

[1]There are in fact two homotopy classes of first order elliptic operators in the complex plane. These are represented by ∂ and $\bar{\partial}$. The null sets of these operators are mappings preserving and reversing orientation respectively (holomorphic or antiholomorphic). Thus the operator S provides a mechanism for us to move from one homotopy component to the other.

2.4. CLASSICAL REGULARITY THEORY

Green's formula gives $\overline{\partial} T\omega = \omega$ and $\partial T\omega = S\omega$ for smooth ω with compact support. The density of smooth functions in L^p and the L^p boundedness of these operators implies that all the above formulas hold whenever $\omega \in L^p(\mathbb{C})$.

Now a solution f to the complex Beltrami equation with compactly supported μ can be found in the form $f(z) = z + T\omega(z)$, where

(2.17) $$\omega = \mu + \mu S\omega \qquad \text{almost everywhere}$$

We shall call f the *principal solution*.

The integral equation at (2.17) is uniquely solved by the Neumann iteration procedure in $L^p(\mathbb{C})$ if

(2.18) $$\|\mu\|_\infty \|S\|_p < 1$$

The fact that it is the invertibility of the Beltrami operator $\mathbf{I} - \mu S : L^p(\mathbb{C}) \to L^p(\mathbb{C})$ which determines the L^p theory of solutions to the Beltrami equation as was first observed by Bojarski, [**21, 22**]. The above representation of the solution f to the Beltrami equation shows that $f \in z + W^{1,p}(\mathbb{C})$ for all p such that $\|\mu\|_\infty \|S\|_p < 1$. Since $\|S\|_2 = 1$ and $\|\mu\|_\infty < 1$ there is always such a $p > 2$. This implies a higher degree of regularity for the principal solution than the initial assumption that it lies in $W^{1,2}_{loc}(\mathbb{C})$. The factorization theorem then shows all $W^{1,2}_{loc}(\mathbb{C})$ solutions enjoy this higher degree of regularity, see [**3, 75, 22**].

This is all clearly explained in Lehto and Virtanen's book [**75**] where they anticipate the optimal results which were finally obtained by Astala [**6**] in spectacular fashion in 1993 using ideas from holomorphic dynamics. Together with previously known and more elementary results, the "Area Distortion Theorem" of Astala can be stated in the following way:

THEOREM 2.1. *Let μ be a measurable function defined in Ω with $\|\mu\|_\infty = k < 1$. Let f be any solution to the Beltrami equation with $f \in W^{1,q}_{loc}(\Omega)$, $q > 1 + k$. Then $f \in W^{1,p}_{loc}(\Omega)$ for all $p < 1 + \frac{1}{k}$. This theorem is sharp; there may be solutions in $W^{1,1+k}_{loc}(\Omega)$ not in any higher Sobolev space, and there may be solutions in $W^{1,2}_{loc}(\Omega)$ not in $W^{1+1/k}_{loc}(\Omega)$.*

Notice the indices p and q in the above result form a Hölder conjugate pair. See [**10**] for a more general discussion and further results in this direction.

Indeed Theorem 2.1 would follow from the methods outlined above if the conjectural values $\|S\|_p = p - 1$ for $p \geq 2$ were to be proven. This perhaps is one of the most important outstanding problem in the planar theory[2].

It is not too difficult to go from Bojarski's representation formula to the existence theorem. The existence theorem for quasiconformal mappings, more recently called the "measurable Riemann mapping theorem", is one of the most fundamental results in the theory and has come to play a central rôle in modern complex analysis, [**87, 19**].

---

[2]There have been important partial results towards the solution of this problem obtained by Volberg and Nazarov [**110**] improving earlier estimates of Banuelos and Wang [**14**] and ourselves [**60**]. The current best estimate is $\|S\|_p \leq 2(p-1)$ for $p \geq 2$.

THEOREM 2.2. *Let μ be a measurable function in a domain $\Omega \subset \mathbb{C}$ and suppose $\|\mu\|_\infty < 1$. Then there is a quasiconformal mapping $g : \Omega \to \mathbb{C}$ whose complex Beltrami coefficient is equal to μ almost everywhere. Moreover every $W^{1,2}_{loc}(\Omega)$ solution f to the Beltrami equation is of the form*
$$f(z) = F(g(z))$$
where $F : g(\Omega) \to \mathbb{C}$ is a holomorphic function.

The main results of this paper are sharp generalizations of these two fundamental theorems in the degenerate elliptic case. That is when $\|\mu\|_\infty = 1$. This follows and extends the pioneering work of David [33]. As discussed in the introduction, there is also related earlier work of Lehto on these more general problems, see [72, 73] where a number of the necessary estimates can be found.

Let us now turn to a more general discussion.

CHAPTER 3

Partial Differential Equations

There is a strong interaction between linear and non-linear elliptic systems in the plane and quasiconformal mappings. The most general first order linear (over the real numbers) elliptic system in the same homotopy class as the classical Cauchy-Riemann operator $\bar{\partial} f = 0$ takes the form
$$\bar{\partial} f = \mu_1 \, \partial f + \mu_2 \, \overline{\partial f}$$
where μ_1 and μ_2 are complex valued measurable functions such that
$$|\mu_1(z)| + |\mu_2(z)| \leq \frac{K-1}{K+1} < 1 \quad a.e. \ \Omega$$
The complex Beltrami equation is simply that equation which is linear over the complex numbers.

These sets of equations are particular cases of the genuine non-linear first order system

(3.1) $$\bar{\partial} f = H(z, \partial f)$$

where $H : \Omega \times \mathbb{C} \to \mathbb{C}$ is Lipschitz in the second variable,
$$|H(z, \zeta) - H(z, \xi)| \leq \frac{K-1}{K+1} |\zeta - \xi|, \quad H(z, 0) \equiv 0$$

The principal feature of the equation at (3.1) is that the difference of two solution need not solve the same equation but it is still K–quasiregular (i.e. it is composition of a quasiconformal homeomorphism and an analytic map). Thus quasiconformal mappings become the central tool in establishing the *a priori* estimates needed for the existence and uniqueness for these non-linear systems of PDEs. We should mention in passing that the non-linear Beltrami system (3.1) also admits quasiconformal solutions $f : \Omega \to \Omega'$ for an arbitrary pair of simply connected domains. These solutions are also unique after suitable normalization. Some results along these lines can be found in [**108, 24, 52**]. A modern approach (that is via harmonic analysis and singular integrals) to these problems can be found in [**8**].

3.1. The Transformation Formula

Here we find it convenient to recall the following useful formula.

Let $f, g : \Omega \hookrightarrow \mathbb{C}$ be quasiconformal mappings. The transformation formula for the Beltrami coefficient of a composition reads as

(3.2) $$\mu_{f \circ g^{-1}}(g(z)) = \frac{\mu_f(z) - \mu_g(z)}{1 - \mu_f(z)\overline{\mu_g(z)}} \left(\frac{\partial g(z)}{|\partial g(z)|} \right)^2$$

An interesting feature here, and which we shall use later, is that due to the change of variables formula the distortion function for f^{-1} may be more regular than that of f. One can see when this might occur from (3.2). Consequently it is possible that the inverse map (when it exists) enjoys substantially better regularity properties than one might *a priori* expect, see for instance §12.2.

3.2. A Fundamental Example

The reader is going to have to bear with us for a moment while we present an example. The importance of the example is that it reflects exactly what is possible in the degenerate elliptic setting, and shows why it is necessary for us to introduce the Orlicz-Sobolev spaces, in the next section, to be able to discuss the fine properties of solutions.

THEOREM 3.1. *Let $\mathcal{A} : [1, \infty) \to [1, \infty)$ be a smooth increasing function with $\mathcal{A}(1) = 1$, $\tau \mathcal{A}'(\tau) \geq 1$ and such that*

$$(3.3) \qquad \int_1^\infty \frac{\mathcal{A}(\tau)}{\tau^2} d\tau < \infty.$$

Then there is a Beltrami coefficient μ compactly supported in the unit disk, $|\mu(z)| < 1$, with the following properties:

(1) *The distortion function*

$$K(z) = \frac{1 + |\mu(z)|}{1 - |\mu(z)|}$$

satisfies

$$(3.4) \qquad \int_{\mathbf{B}} e^{\mathcal{A}(K(z))} dz < \infty$$

(2) *The only $W^{1,1}_{loc}(\mathbf{B})$-solutions to the Beltrami equation*

$$(3.5) \qquad f_{\bar{z}} = \mu f_z \qquad a.e. \quad \mathbf{B}$$

which are continuous at the origin are the constant functions.

(3) *There is a bounded solution $w = f(z)$ to the Beltrami equation in the space weak-$W^{1,2}(\mathbf{B}) \subset \bigcap_{1 \leq q < 2} W^{1,q}(\mathbf{B})$ which homeomorphically maps the punctured disk $\mathbf{B} \setminus \{0\}$ onto the annulus $1 < |w| < R$.*

A few remarks. First, $W^{1,1}_{loc}(\mathbf{B})$ is really the smallest space in which one can begin to discuss what it means to be a solution. Secondly, the integrability condition (3.3) implies that $\mathcal{A}$ is slightly more than sublinear. For instance the function, defined for large values of τ by

$$(3.6) \qquad \mathcal{A}(\tau) = \frac{\tau}{(\log \tau)^{1+\epsilon}}$$

satisfies (3.3) for all $\epsilon > 0$, but not for $\epsilon = 0$. More generally, if we put

$$\log_1 \tau = \log \tau, \qquad \log_{n+1} \tau = \log(\log_n \tau)$$

we have the *iterated logarithm* functions. Then for each $n > 0$ the function

$$(3.7) \qquad \mathcal{A}(\tau) = \frac{\tau}{\log_1(1+\tau) \log_2(e+\tau) \log_3(e^e + \tau) \cdots (\log_n(e^{e^{\cdot^{\cdot^{\cdot}}}} + \tau))^{1+\epsilon}}$$

again satisfies (3.3) for all $\epsilon > 0$, but not for $\epsilon = 0$. Next, the condition $s\mathcal{A}'(s) \geq 1$ is merely a technical assumption to simplify a calculation below. Finally, (3) is no accident. We shall show later that if $e^{\lambda K} \in L^1_{loc}(\mathbf{B})$ for some $\lambda > 0$, then there is a homeomorphic (and hence continuous) solution in $W^{1,p}(\mathbf{B})$ for all $p < 2$. Indeed we shall prove much more. We prove existence of homeomorphic solutions for subexponentially integrable distortion for certain $\mathcal{A}$ as above but with the integral at (3.3) diverging. We also prove the solution lies in an Orlicz class slightly below $W^{1,2}_{loc}$ and that we can also see the jump in regularity of the differential to an Orlicz class slightly above $W^{1,2}_{loc}$.

3.3. The Construction

Given a function $\mathcal{A}$ satisfying (3.3) we define a function $K(s) \geq 1$, for $0 < s \leq 1$, via the functional relation

$$(3.8) \qquad K(s)e^{\mathcal{A}(K(s))} = \frac{e}{s^2}$$

As the map $K \mapsto Ke^{\mathcal{A}(K)}$ is monotone increasing we see that the solution $K(s)$ is well defined and unique with $K(1) = 1$. In fact the solution is decreasing and

$$(3.9) \qquad K'(s) = -\frac{2K(s)}{s(1 + K(s)\mathcal{A}'(K(s)))} \geq -\frac{K(s)}{s}$$

as $K(s)\mathcal{A}'(K(s)) \geq 1$. Hence

$$\frac{d}{dt}(tK(t)) \geq 0$$

and so upon integration from s to 1 we obtain

$$(3.10) \qquad K(s) \leq \frac{1}{s}.$$

Then

$$e^{\mathcal{A}(K(s))} = \frac{e}{s^2 K(s)} \geq \frac{e}{s}$$

$$K(s) \geq \mathcal{A}^{-1}(\log \frac{e}{s})$$

From this we deduce that for all $0 < t \leq 1$

$$\int_0^t \frac{ds}{sK(s)} \leq \int_0^t \frac{ds}{s\mathcal{A}^{-1}(\log e/s)}$$
$$= \int_a^\infty \frac{\mathcal{A}'(\tau)d\tau}{\tau}$$
$$\leq \int_1^\infty \frac{\mathcal{A}'(\tau)d\tau}{\tau}$$

where $a = \mathcal{A}^{-1}(\log \frac{e}{t}) \geq \mathcal{A}^{-1}(1) = 1$. The next lemma, which follows from integration by parts and a minor estimate, asserts this last integral is finite.

LEMMA 3.2. *If $\mathcal{A}$ is an increasing positive function, then*

$$\int_1^\infty \frac{\mathcal{A}(\tau)\,d\tau}{\tau^2} < \infty \quad \text{if and only if} \quad \int_1^\infty \frac{\mathcal{A}'(\tau)\,d\tau}{\tau} < \infty$$

Thus

$$\int_0^t \frac{ds}{sK(s)}$$

is a well defined bounded increasing function. Next

$$\int_{\mathbf{B}} e^{\mathcal{A}(K(|z|))} \, dz = 2\pi \int_0^1 e^{\mathcal{A}(K(s))} s \, ds$$

$$= 2\pi e \int_0^1 \frac{ds}{sK(s)}$$

$$\leq \int_1^\infty \frac{\mathcal{A}'(\tau) d\tau}{\tau}$$

So that the function K is subexponentially integrable. Now set

(3.11) $$f(z) = \frac{z}{|z|} \rho(|z|)$$

where

(3.12) $$\rho(t) = \exp\left(\int_0^t \frac{ds}{sK(s)}\right)$$

Then f is a radial stretching, defined on $\mathbf{B} \setminus \{0\}$, and we compute (away from the origin)

$$\dot{\rho}(t) = \frac{\rho(t)}{tK(t)}$$

$$K(z,f) = \frac{\rho}{|z|\dot{\rho}} = K(|z|)$$

$$|Df(z)| = \frac{\rho(|z|)}{|z|}$$

$$J(z,f) = \frac{\rho^2(|z|)}{K(z)|z|^2}$$

$$\mu_f(z) = -\frac{z}{\bar{z}} \frac{K(z) - 1}{K(z) + 1}$$

Notice here that the function f is not continuous at the origin. However, near the origin we have $\rho(|z|) \approx 1$ and therefore the formula for $|Df|$ gives us the bounds

(3.13) $$\frac{1}{|z|} \leq |Df(z)| \leq \frac{C}{|z|}$$

which uncovers our bounded solution in weak-$W^{1,2}$.

Notice that $w = f(z)$ is a C^∞ diffeomorphism in the punctured disk onto the annulus $1 < |w| < R = \rho(1)$; and we have "cavitation" at the origin.

Finally we want to show that there are no nonconstant continuous solutions in $W^{1,1}_{loc}(\mathbf{B})$. To this end let $\epsilon > 0$ and set $r = \rho(\epsilon)$. Note that $r \to 1$ as $\epsilon \to 0$. Let g be a $W^{1,1}_{loc}(\mathbf{B}) \cap C(\mathbf{B})$ solution to (3.5) with $\mu = \mu_f$ as above. Set

$$\varphi(z) = g \circ f^{-1} : \{r < |w| < R\} \to g(\{\epsilon < |z| < 1\})$$

As $\varphi \in W^{1,1}(\{r < |w| < R\})$, a simple computation shows $\bar{\partial}\varphi = 0$ and the Weyl lemma gives φ holomorphic. This is true for every $\epsilon > 0$ and so for every $r > 1$ and thus we are provided with an analytic function for which

$$\varphi \circ f(z) = g(z), \quad z \in \mathbf{B} \setminus \{0\}$$

This equation implies the function $\varphi = \varphi(w)$ has the limit $g(0)$ as $|w| \to 1$. Hence both φ and g are constant. $\square$

3.4. Cavitation and Riemann Surfaces

The above example was partly motivated by an earlier result of Ball, [**12**]. The example presented above seems to be optimal with respect to the property of cavitation, a phenomena much studied in non-linear elasticity. This is where a map not only fails to be discontinuous, but "holes" appear in the image. The example suggests the possible extremes and raises the following:

Let $\mathcal{A}$ be a smooth increasing real valued function such that (3.3) fails, so

$$\int_1^\infty \mathcal{A}(\tau) \frac{d\tau}{\tau^2} = \infty \tag{3.14}$$

and let μ be a Beltrami coefficient which is compactly supported in the unit ball and whose distortion function $K(z) = \frac{1+|\mu(z)|}{1-|\mu(z)|}$ satisfies

$$\int_{\mathbf{B}} e^{\mathcal{A}(K(z))} \, dz < \infty \tag{3.15}$$

Our question is then: Is any solution to the Beltrami equation with $|Df| \in$ weak-$W^{1,2}(\mathbf{B})$ is continuous ? The answer is certainly YES if we consider solutions with $|Df| \in W^{1,2}(\mathbf{B})$.

Notice that the map described above takes the punctured disk $\{0 < |z| < 1\}$ homeomorphically onto the conformally distinct annulus $\{1 < |z| < R\}$. The hyperbolic area of $\{0 < |z| < \frac{1}{2}\}$ in the punctured disk is finite, whereas that of its image is clearly infinite in the hyperbolic metric. Such maps can be pieced together in an obvious way to provide maps of analytically finite Riemann surfaces (roughly finite area and finite topology, for instance $F \setminus \Sigma$ with F a closed Riemann surface Σ a finite set, see [**82**]) to a homeomorphic open Riemann surface which is not analytically finite. One might ask the question:

Let $\mathcal{A}$ be a smooth increasing real valued function satisfying (3.14) and let F_1, F_2 be Riemann surfaces with F_1 analytically finite. Suppose $f : F_1 \to F_2$ is a mapping (say in weak-$W^{1,2}_{loc}$) with

$$\int_{F_1} e^{\mathcal{A}(K(z))} \, dz < \infty$$

Then F_2 is analytically finite.

Notice that as soon as F_2 is analytically finite, the two surfaces are quasiconformally equivalent. For more general types of surfaces, for instance those with "infinite" topology, one needs a substantially different type of result. In particular one needs details of the size of the removable sets for such mappings. A first attempt at this problem is a subject of our joint work with K. Astala and P. Koskela, [**7**] and is recounted in §19 of this paper.

Such results would give limitations on potential developments and extensions of Teichmüller theory. Moreover, there are clear applications in holomorphic dynamics. It is a simple consequence of our results as presented here, that the question raised above has an affirmative answer if $\mathcal{A}(t) = t/\log(e+t)$, even with a slightly lesser regularity assumption, see §§11.2, 12.2 and 13.2.

CHAPTER 4

Mappings of Finite Distortion

We shall now give a general definition of the sort of mappings which we will mostly consider in this paper. These are the mappings of finite distortion.

Definition A mapping $f : \Omega \to \mathbb{C}$ is said to have *finite distortion* if:
(1) $f \in W^{1,1}_{loc}(\Omega)$,
(2) The Jacobian determinant of f is locally integrable and does not change sign in Ω
(3) There is a measurable function $\mathcal{K} = \mathcal{K}(z) \geq 1$, finite almost everywhere, such that f satisfies the *distortion inequality*

(4.1) $\qquad\qquad |Df(z)|^2 \leq \mathcal{K}(z) |J(z,f)| \qquad$ a.e. Ω

Notice that the hypotheses are not sufficient to guarantee that $f \in W^{1,2}_{loc}(\Omega)$ unless the distortion function $\mathcal{K}$ is bounded. Nor do they imply that the Jacobian does not vanish on a set of positive measure, see [**59**] for an example. Further, we note that condition 2. above is automatically satisfied for a homeomorphism of Sobolev class $W^{1,1}(\Omega)$. Also note that if f is a homeomorphism then the Gehring-Lehto Theorem [**40**] shows f to be differentiable almost everywhere. In general is not known whether the Jacobian of a homeomorphism $f : \Omega \subset \mathbb{R}^n \to \Omega' \subset \mathbb{R}^n$ of class $W^{1,p}_{loc}(\Omega, \Omega')$ with $1 \leq p < n-1$ may change sign.

A recent important and remarkable result of Hencl and Koskela, [**49**] Theorem 1.2, states that if $f : \Omega \to \Omega'$ is a planar homeomorphism of finite distortion, then $f^{-1} : \Omega' \to \Omega$ is a mapping of finite distortion. Of course in general the composition of mappings of finite distortion will not have finite distortion.

Of course an orientation preserving mapping of finite distortion has an associated Beltrami coefficient

$$\mu(z) = \frac{f_{\bar{z}}(z)}{f_z(z)}$$

and will therefore be a very weak solution to a Beltrami equation. However, general mappings of finite distortion whose differential is in a good integrability class enjoy some nice properties which we shall need from time to time. Principal among these is monotonicity.

Conversely, a $W^{1,1}_{loc}$ solution to a Beltrami equation is a mapping of finite distortion provided that $|\mu(z)| < 1$ almost everywhere and 2. above holds (eg. a homeomorphic solution). Thus the concept of a mapping of finite distortion will allow us to discuss properties of fairly general solutions to the Beltrami equation.

Indeed, we may write the Beltrami equation in real variables as the nonlinear first order PDE

(4.2) $\qquad\qquad D^t f(z) D f(z) = J(z,f) G(z), \qquad z \in \Omega \subset \mathbb{R}^2$

where $G(z)$ is a measurable function valued in the space of symmetric positive definite 2×2 matrices of determinant 1. We have the pointwise almost everywhere estimate

$$\frac{1}{\mathcal{K}(z)}|\zeta|^2 \leq \langle G(z)\zeta, \zeta \rangle \leq \mathcal{K}(z)|\zeta|^2 \tag{4.3}$$

for vectors $\zeta \in \mathbb{R}^2$ and thus the distortion function $\mathcal{K}$ provides ellipticity bounds for the equation. The classical case of $\mathcal{K}$ bounded, gives uniform ellipticity estimates on G.

The theory of mappings of finite distortion, both in the plane and in higher dimensions where the definition is much the same, has elicited a great deal of recent study. In fact these mappings are a focus of the monograph [58]. Other recent advances appear in [7, 54, 68, 69, 70] and in many other places. The reader will see that from time to time the results we offer here will generalize to higher dimensions, though there are usually some complications. However, oftentimes there is no clear path to establishing a higher dimensional analogue and so there remain a number of interesting questions.

4.1. Orlicz–Sobolev Spaces

The examples of the previous section shows that if we are to establish the existence of continuous, and in particular homeomorphic, solutions to the Beltrami equation in the degenerate elliptic case we are going to have to look in spaces that are very close to $W^{1,2}$. Moreover in order to extract the full benefit from our studies of mappings of finite distortion it is necessary to have at hand rather more sophisticated techniques than those used in the classical setting. This brings us to the Orlicz–Sobolev spaces and it is the purpose of this section to introduce the basic notions and establish a few basic properties of these spaces.

An *Orlicz* function is a continuously increasing function

$$P : [0, \infty) \to [0, \infty), \quad P(0) = 0, \quad \lim_{t \to \infty} P(t) = \infty, \tag{4.4}$$

though in all our applications P will usually be convex. The Orlicz space $L^P(\Omega, \mathbb{V})$ consists of those Lebesgue measurable mappings f defined in Ω and valued in the finite dimensional inner-product space $\mathbb{V}$ such that

$$\int_\Omega P(\lambda|f|) < \infty, \quad \text{for some } \lambda = \lambda(f) > 0. \tag{4.5}$$

This is a complete linear metric space with respect to the L^P distance defined by

$$\operatorname{dist}_P(f, g) = \inf \left\{ \frac{1}{\lambda} : \lambda \int_\Omega P(\lambda|f - g|) \leq 1 \right\} \tag{4.6}$$

We shall also make use of the non-linear *Luxemburg functional*

$$\|f\|_P = \inf \left\{ \frac{1}{\lambda} : \int_\Omega P(\lambda|f|) \leq P(1) \right\} \tag{4.7}$$

which is homogeneous, but in general fails to satisfy the triangle inequality. A convex Orlicz function P is often called a *Young* function. In this case $\|\cdot\|_P$ is known to be a norm and $L^P(\Omega, \mathbb{V})$ with this norm is a Banach space. As a first example, if we put $P(t) = t^p$, $0 < p < \infty$ then the spaces $L^P(\Omega, \mathbb{V})$ coincide with the usual Lebesgue spaces $L^p(\Omega, \mathbb{V})$. Note that L^p is a Banach space only when

$p \geq 1$. The basic examples that we have in mind are the *Zygmund spaces*, denoted $L^p \log^\alpha L(\Omega, \mathbb{V})$, corresponding to the Orlicz function $P(t) = t^p \log^\alpha(a+t)$ with $1 \leq p < \infty$, $\alpha \in \mathbb{R}$ and suitably large constant a.

As an example the defining function $P(t) = t^p \log^\alpha(e+t)$, $1 \leq p < \infty$, is a Young function when $\alpha \geq 1 - p$, and there we have the following straightforward estimates;

$$\|f\|_{L^p \log^{-1} L} \leq \|f\|_p \leq \|f\|_{L^p \log L} \tag{4.8}$$

and

$$\|f\|_{L^p \log L} \leq \left[\int |f|^p \log \left(e + \frac{|f|}{\|f\|_p} \right) \right]^{\frac{1}{p}} \leq 2 \|f\|_{L^p \log L} \tag{4.9}$$

As a matter of fact the integral expression in (4.9) defines a norm in $L^p \log L$, though the triangle inequality is far from obvious. Another important example which will occur frequently is the exponential class $Exp(\Omega)$ defined with the Orlicz function $P(t) = e^t - 1$, or $Exp_\alpha(\Omega)$ with $P(t) = e^{t^\alpha} - 1$, $\alpha > 0$.

We will need to consider various dual spaces and to do this we really need to confine our attention to Orlicz functions represented by the integral

$$P(t) = \int_0^t \alpha(s) \frac{ds}{s} \tag{4.10}$$

where we assume that α is a positive smooth function on $\mathbb{R}_+$ with $\alpha'(s) > 0$ and such that the integral at (4.10) converges. This time $\alpha(t) = t^p$, $0 < p < \infty$, leads to the usual Lebesgue spaces. Note here however, that the inverse function α^{-1} also meets all these requirements, provided the corresponding integral converges.

Aspects of the basic theory of Orlicz functions can be found in [**95**], here we recount a few of the more central results which will be important to us.

Given a collection $\alpha_1, \alpha_2, \ldots, \alpha_k$ of such functions we define α such that

$$\alpha^{-1}(t) = \alpha_1^{-1}(t) \cdot \alpha_2^{-1}(t) \cdots \alpha_k^{-1}(t). \tag{4.11}$$

We define the corresponding Orlicz functions

$$P_i(t) = \int_0^t \alpha_i(s) \frac{ds}{s}, \quad i = 1, 2, \ldots, k, \qquad P(t) = \int_0^t \alpha(s) \frac{ds}{s}. \tag{4.12}$$

The functions have the important property of satisfying Young's inequality.

$$P(t_1 \cdot t_2 \cdots t_k) \leq P_1(t_1) + P_2(t_2) + \cdots + P_k(t_k) \tag{4.13}$$

The reader may care to consider the classical setting where $\alpha_i(t) = t^{p_i}$, with $p_1, p_2, \ldots, p_k > 0$ in which case $\alpha(t) = t^p$ where $\frac{1}{p} = \frac{1}{p_1} + \frac{1}{p_2} + \cdots \frac{1}{p_k}$. We also leave it to the reader to use (4.13) to prove the following analogue of Hölder's inequality:

$$\|\varphi_1 \varphi_2 \cdots \varphi_k\|_P \leq \|\varphi_1\|_{P_1} \|\varphi_2\|_{P_2} \cdots \|\varphi_k\|_{P_k} \tag{4.14}$$

for a collection of functions $\varphi_i \in L^{P_i}(\Omega)$.

Hölder's inequality for Zygmund spaces will be quite important to us. It takes the form

$$\|\varphi_1 \cdots \varphi_k\|_{L^p \log^\alpha L} \leq C \|\varphi_1\|_{L^{p_1} \log^{\alpha_1} L} \cdots \|\varphi_k\|_{L^{p_k} \log^{\alpha_k} L} \tag{4.15}$$

where $p_1, p_2, \ldots, p_k > 1$, $\alpha_1, \alpha_2, \ldots, \alpha_k \in \mathbb{R}$ and
$$\frac{1}{p} = \frac{1}{p_1} + \frac{1}{p_2} + \cdots + \frac{1}{p_k}, \qquad \frac{\alpha}{p} = \frac{\alpha_1}{p_1} + \frac{\alpha_2}{p_2} + \cdots + \frac{\alpha_k}{p_k}.$$
The constant here does not depend on the functions $\varphi_i \in L^{p_i} \log^{\alpha_i} L$.

A pair of Orlicz functions (P, Q) are called a Hölder conjugate couple, or *Young complementary functions*, if we have Hölder's inequality

(4.16) $$\left| \int_\Omega \langle f, g \rangle \right| \leq C_{PQ} \|f\|_P \|g\|_Q$$

for $f \in L^P(\Omega, \mathbb{V})$ and $g \in L^Q(\Omega, \mathbb{V})$.

Our basic example is the Hölder conjugate couple $P(t) = t \log(e + t)$ and $Q(t) = e^t - 1$ defining the Zygmund and exponential classes respectively. In this case we have the estimate

(4.17) $$\left| \int_\Omega \langle f, g \rangle \right| \leq 4 \|f\|_{L \log L} \|g\|_{Exp}$$

Indeed in view of the same homogeneities on each side we can assume that each of the Luxemburg norms is equal to 1. From the definition of these norms we find
$$\int_\Omega |f| \log(e + |f|) = \log(e + 1) \quad \text{and} \quad \int_\Omega (e^{|g|} - 1) = e - 1$$
Then we use the elementary inequality

(4.18) $$|f||g| \leq |f| \log(1 + |f|) + e^{|g|} - 1$$

to conclude that $\int_\Omega |f||g| \leq 4$, as required.

Having introduced Orlicz spaces, we now turn to Orlicz–Sobolev spaces. For an Orlicz function P the space $W^{1,P}(\Omega, \mathbb{V})$ can be defined in much the same way as in the classical case $P(t) = t^p$. In order to speak of the distributional derivatives, however, it is of course necessary that functions in $L^P(\Omega, \mathbb{V})$ are at least locally integrable. This forces upon us the assumption that for all sufficiently large t,

(4.19) $$P(t) \geq a\, t, \qquad \text{for some } a > 0.$$

Under this assumption we make the following definition.

Definition A distribution $f \in \mathcal{D}'(\Omega, \mathbb{V})$ belongs to the Orlicz–Sobolev space $W^{1,P}(\Omega, \mathbb{V})$ if the partials f_z and $f_{\bar{z}}$ are represented by functions in $L^P(\Omega, \mathbb{V})$. Here $\mathcal{D}'(\Omega, \mathbb{V})$ is the space of Schwartz distributions.

In all of our subsequent notations we will omit reference to the target space $\mathbb{V}$ if it plays no fundamental role in the results.

It is apparent that many of the basic notions and results in the theory of Sobolev spaces carry over to this more general setting without any difficulty. Some however, lead to significant generalizations and new and indispensable tools in non-linear analysis. To illustrate this we recall the Imbedding Theorem which asserts that $W^{1,n}(\Omega, \mathbb{V}) \stackrel{i}{\hookrightarrow} Exp(\Omega)$, though actually the stronger result of Trudinger holds [41]

(4.20) $$W^{1,n}(\Omega) \stackrel{i}{\hookrightarrow} Exp_{\frac{n}{n-1}}(\Omega)$$

This added degree of regularity is essential in the theory of PDEs . We too shall see how Orlicz–Sobolev spaces can be used to find a little, but very important, improvement in regularity.

One effect of using Orlicz–Sobolev spaces is monotonicity and modulus of continuity estimates. As these will be of fundamental importance to us later we turn briefly to consider these ideas.

4.2. Monotonicity

In one-dimension a function $u : \Omega \to \mathbb{R}$ is monotone if it satisfies both a maximum and minimum principle on every subinterval. Equivalently, we have the oscillation bounds $\operatorname{osc}_I u \leq \operatorname{osc}_{\partial I} u$ for every interval $I \subset \Omega$. The definition of monotonicity in higher dimensions closely follows this observation.

A continuous function $u : \Omega \to \mathbb{R}$ defined in a domain Ω is *monotone* if

$$\operatorname{osc}_{\mathbf{B}} u \leq \operatorname{osc}_{\partial \mathbf{B}} u \tag{4.21}$$

for every ball $\mathbf{B} \subset \Omega$. This definition in fact goes back to Lebesgue in 1907 where he first showed the relevance of the notion of monotonicity to elliptic PDEs in the plane. In order to handle very weak solutions of various differential inequalities, such as the distortion inequality, we need to extend this concept, dropping the assumption of continuity, and to the setting of Orlicz–Sobolev spaces.

Definition. A real valued function $u \in W^{1,P}(\Omega)$ is said to be *weakly monotone* if for every ball $\mathbf{B} \subset \Omega$ and all constants $m \leq M$ such that

$$|M - u| - |u - m| + 2u - m - M \in W^{1,P}_0(\mathbf{B}) \tag{4.22}$$

we have

$$m \leq u(x) \leq M \tag{4.23}$$

for almost every $x \in \mathbf{B}$. Here $W^{1,P}_0(\mathbf{B})$ consists of those functions in $W^{1,P}(\mathbf{B})$ which vanish on the boundary of $\mathbf{B}$ in the Sobolev sense, see [58].

For continuous functions (4.22) holds if and only if $m \leq u(x) \leq M$ on $\partial \mathbf{B}$. Then (4.23) says we want the same condition in $\mathbf{B}$, that is the maximum and minimum principles. Precisely we have

LEMMA 4.1. *Let Ω be a bounded domain and suppose that $u \in W^{1,P}(\Omega) \cap C(\overline{\Omega})$ is weakly monotone. Then*

$$\min_{\partial \Omega} u \leq u(x) \leq \max_{\partial \Omega} u \tag{4.24}$$

for every $x \in \Omega$.

Proof. Another way to look at condition (4.22) is that both $(u-m)^-$ and $(u-M)^+$ belong to $W^{1,P}_0(\mathbf{B})$, where the negative and positive parts of v are defined by the equation $2v^\pm = |v| \pm v$. Thus (4.24) follows by applying (4.23) to suitable truncations, say $u_{\pm\epsilon}(x) = u(x)$ on the open subregion $\Omega_{\pm\epsilon} = \{x : \pm u(x) \geq \pm\epsilon\}$ for both choices of sign. $\square$

The paper [79] by Manfredi should be mentioned as the beginning of the systematic study of weakly monotone functions and perhaps he was the first to introduce this notion.

We now recall a fundamental monotonicity result in the Orlicz–Sobolev classes, Theorem 6.3.1 [58]. First we need to discuss the class of Orlicz functions for which our method applies.

4.3. A Class of Orlicz Functions

Let $\varphi \in C^1(0,1]$ be any positive decreasing function with $\lim_{s\to 0} \varphi(s) = \infty$. We define

$$(4.25) \qquad \Phi(t) = \sup_{0 < \epsilon < 1} \frac{-1}{\varphi(\epsilon)} \int_\epsilon^1 st^{2-s} d\varphi(s)$$

The Orlicz functions we consider need only satisfy the lower bound $P(t) \geq c\, \Phi(t)$ for some positive constant c. We record this condition as

$$(4.26) \qquad \varphi(\epsilon) P(t) \geq -c \int_\epsilon^1 st^{2-s} d\varphi(s)$$

for all $0 < \epsilon < 1$ and all $t \geq 1$.

Note that Φ is increasing and convex and satisfies

$$(4.27) \qquad \int_1^\infty \Phi(t) \frac{dt}{t^3} = \infty,$$

and hence

$$\int_1^\infty P(t) \frac{dt}{t^3} = \infty$$

Indeed for all $0 < \epsilon \leq 1$ we have

$$-\frac{1}{\varphi(\epsilon)} \int_\epsilon^1 \frac{s\, d\varphi(s)}{t^{1+s}} \leq \frac{\Phi(t)}{t^3}$$

Fixing an arbitrary $a \geq 1$ and integrating this with respect to t from a to ∞ we obtain,

$$-\frac{1}{\varphi(\epsilon)} \int_\epsilon^1 \frac{d\varphi(s)}{a^s} \leq \int_a^\infty \frac{\Phi(t)}{t^3} dt$$

Then for every δ, $\epsilon \leq \delta < 1$, we can write

$$\frac{\varphi(\epsilon) - \varphi(\delta)}{a^\delta \varphi(\epsilon)} \leq -\frac{1}{\varphi(\epsilon)} \int_\epsilon^1 \frac{d\varphi(s)}{a^s} \leq \int_a^\infty \frac{\Phi(t)}{t^3} dt$$

Finally, letting $\epsilon \to 0$ we have for all $0 < \delta \leq 1$,

$$\frac{1}{a^\delta} \leq \int_a^\infty \frac{\Phi(t)}{t^3} dt$$

We now pass to the limit as $\delta \to 0$ to obtain

$$1 \leq \int_a^\infty \frac{\Phi(t)}{t^3} dt$$

for all $a \geq 1$. This is only possible if the integral at (4.27) diverges.

In fact the reader may care to verify that if we put $\varphi(s) = \frac{1}{s}$ we get a function Φ for which

$$\frac{c_1 t^2}{\log(1+t)} \leq \Phi(t) \leq \frac{c_2 t^2}{\log(1+t)}$$

However, for our purposes a quite satisfactory Orlicz function is obtained by setting

$$\varphi(s) = \log \frac{e}{s}, \quad \text{with } 0 < s \leq 1$$

and which give the bounds

$$\text{(4.28)} \qquad \frac{c_1\, t^2}{\log(e+t)\log\log(e+t)} \leq \Phi(t) \leq \frac{c_2\, t^2}{\log(e+t)\log\log(e+t)}$$

for all $t \geq 0$ where the constants are positive, see [56, 58].

It is worth pointing out here that the divergence condition at (4.27), modulo few minor technical assumptions on P, is also sufficient to obtain the lower bound at (4.26). This is an elementary fact but requires rather lengthy calculations and a fine analysis of the behaviour of certain integrals. We have postponed this calculation to Chapter 8 - Some Technical Results - where we discuss the important implications of this observation. The reader will see that the bound at (4.26) is the crucial fact used in our proofs.

4.4. The Monotonicity Theorem

THEOREM 4.2. *Let $P(t)$ be an Orlicz function such that the bound at (4.26) holds. Then the coordinate functions of mappings with finite distortion in $W^{1,P}(\Omega)$ are weakly monotone. In particular, if*

$$\text{(4.29)} \qquad P(t) \geq \frac{C\, t^2}{\log(e+t)\log\log(3+t)}.$$

then the conclusion remains valid.

We do not prove this theorem here as it would lead us too far astray. However, we point out to the reader that this is a deep result with fairly profound consequences. One might think of it as a generalization of the maximum principle. A complete proof is given in [58] Chapter 7. Very recent generalizations of these sorts of results for mappings between Riemannian manifolds can be found in [47].

There is a particularly elegant geometric approach to the continuity estimates of monotone functions. The idea goes back to Gehring [37] in his study of the Liouville theorem in space. While many other interesting applications of Gehring's oscillation lemma have been discussed in the literature, its use with weakly monotone functions seems less familiar.

LEMMA 4.3. *Let $u \in W^{1,p}(\mathbf{B}_R)$ with $p > 1$, be weakly monotone in the ball $\mathbf{B}_R = \mathbf{B}(x_0, R)$. For all Lebesgue points $a, b \in \mathbf{B}_r \subset \mathbf{B}_R$, we have*

$$\text{(4.30)} \qquad |u(a) - u(b)| \leq \pi\, t \left(\frac{1}{2\pi t} \int_{\partial \mathbf{B}_t} |\nabla u|^p \right)^{\frac{1}{p}}$$

for almost every $t \in [r, R]$.

The numbers $t \in [r, R]$ for which (4.30) holds can be identified as the Lebesgue points of the function $t \mapsto \int_{\partial \mathbf{B}_t} |\nabla u|^p$ which is certainly an integrable function on the interval $[r, R]$ by Fubini's Theorem.

4.5. Modulus of Continuity

We need to make the standing condition that both (4.25) and (4.26) are valid for the Orlicz function P.

In order to present other explicit bounds, we need to introduce the *P–modulus of continuity* $\Xi_P(\tau)$ defined for $0 \leq \tau < 1$ as follows. For $\tau > 0$ the value t of Ξ_P at τ is uniquely determined by the equation

$$\int_1^{1/\tau} P(st)\,\frac{ds}{s^3} = P(1). \tag{4.31}$$

Certainly Ξ_P is a non-decreasing function and condition (4.27) ensures that

$$\lim_{\tau \to 0} \Xi_P(\tau) = 0. \tag{4.32}$$

Thus we may legitimately define $\Xi_P(0) = 0$. We shall henceforth drop the subscript P from Ξ_P when P is understood from the context.

In addition to the condition (4.27) we will also make one more rather technical requirement:

The function $s \mapsto P(s^{\frac{1}{\gamma}})$ is convex for some exponent $\gamma > 1$.

We call γ, the *exponent of convexity*. This requirement will prove no obstacle in what follows since we will be confining ourselves to Orlicz functions close to the natural one $P(t) = t^2$, so that in fact in all our applications the convexity will hold for any $1 < \gamma < 2$ and all sufficiently large t.

Given the transcendental nature of the equation one must solve, it is impossible in all but the most elementary situations, to calculate Ξ. Hence we next give explicit formulas for $\Xi(\tau)$ which exhibit the correct asymptotics for τ near 0. First the exact formula

$$P(t) = t^2, \quad \Xi(\tau) = |\log \tau|^{-\frac{1}{2}} \tag{4.33}$$

More generally for all $\alpha > 0$ we have,

$$P(t) = t^2 \log^{\alpha-1}(e+t),\ \alpha > 0, \quad \Xi(\tau) \approx |\log \tau|^{-\frac{\alpha}{2}} \tag{4.34}$$

$$P(t) = \frac{t^2}{\log(e+t)}, \quad \Xi(\tau) \approx [\log|\log \tau|]^{-\frac{1}{2}}, \tag{4.35}$$

and finally

$$P(t) = \frac{t^2}{\log(e+t)\log\log(3+t)}, \quad \Xi(\tau) \approx [\log\log|\log \tau|]^{-\frac{1}{2}}, \tag{4.36}$$

We now have the fundamental modulus of continuity estimate Theorem 6.5.1 [58].

THEOREM 4.4. *Let $u \in W^{1,P}(\mathbf{B})$ be weakly monotone in $\mathbf{B} = \mathbf{B}(z_0, 2R)$, where P has exponent of convexity equal to γ. Then for all Lebesgue points $a, b \in \mathbf{B}(z_0, R)$ we have*

$$|u(a) - u(b)| \leq \frac{4\pi \gamma R}{\gamma - 1} \|\nabla u\|_{\mathbf{B},P}\, \Xi\left(\frac{|a-b|}{2R}\right). \tag{4.37}$$

In particular, u has a continuous representative for which (4.37) holds for all a and b in the disk $\mathbf{B}(z_0, R)$.

In the statement above we have used

(4.38) $$\|\nabla u\|_{\mathbf{B},P} = \inf\left\{\frac{1}{\lambda} : \frac{1}{|\mathbf{B}|}\int_{\mathbf{B}} P(\lambda|\nabla u|) \leq P(1)\right\}$$

to denote the P-average of ∇u over the ball $\mathbf{B}$. So for example, if $P(t) = t^p$, $p > 1$, then

$$\|\nabla u\|_{\mathbf{B},P} = \left(\frac{1}{|\mathbf{B}|}\int_{\mathbf{B}} |\nabla u|^p\right)^{\frac{1}{p}}$$

We conclude this section with the following basic result, [**56, 58**],

THEOREM 4.5. *Every mapping with finite distortion in the Orlicz–Sobolev class $W^{1,P}_{loc}(\Omega)$, with P satisfying (4.29), is continuous.*

Based on the results presented in the Appendices, this theorem remains valid for Orlicz functions P satisfying the divergence condition at (4.27) modulo certain technical assumptions discussed there, namely those of Theorem 8.1. For other recent advances in this area we refer the reader to [**69, 70**] and for earlier relevant work the reader should consult [**42, 109, 80, 81, 48, 86, 77, 68**].

CHAPTER 5

Hardy Spaces and BMO

Delicate cancellation properties of various non-linear objects such as Jacobians cannot be discussed without introducing the Hardy spaces, and in order to fully understand the cancellation phenomenon one must work with the Hardy-Orlicz type spaces. It is our objective here to give a brief account of this.

We are concerned with the Hardy-Orlicz spaces $H^P(\Omega)$ on domains $\Omega \subset \mathbb{C}$, where P is a fairly general Orlicz function. These spaces have already appeared under the name generalized Hardy spaces in the work of Janson [66] in 1980. Our definition follows closely the maximal characterization of the classical Hardy spaces $H^p(\mathbb{C})$ with $0 < p \leq 1$. That is, we first define a maximal function of a distribution and then assume that this function belongs to a suitable Orlicz space.

5.1. Mollifiers

We shall rely on particular mollifiers constructed as follows. Fix a nonnegative function $\Phi \in C_0^\infty(\mathbb{C})$ supported in the closed unit ball and having integral 1. For example

$$\begin{aligned}(5.1) \qquad \Phi(z) &= C \exp\left(\frac{1}{|z|^2 - 1}\right) && \text{if } |z| < 1 \\ \Phi(z) &= 0 && \text{if } |z| \geq 1\end{aligned}$$

where the constant C is chosen so that $\int \Phi(z)\,dz = 1$. Now set

$$(5.2) \qquad \Phi_t(x) = t^{-2}\Phi(t^{-1}x), \qquad t > 0$$

It is easy to see that

$$(5.3) \qquad \lim_{t \to 0} \Phi_t(a - x) = \delta_a$$

in the sense of distributions, that is

$$(5.4) \qquad \lim_{t \to 0} \int \Phi_t(a - x)\varphi(x)\,dx = \varphi(a) = \delta_a[\varphi]$$

for every test function φ.

The integral on the left hand side is called the *convolution* of φ with the mollifier Φ_t, usually denoted by $(\varphi * \Phi_t)(a)$. Less obvious, but of crucial importance is that for every $f \in L^1_{loc}(\mathbb{C}, \mathbb{V})$ (the locally integrable functions defined on $\mathbb{C}$ and valued in some finite dimensional vector space)

$$(5.5) \qquad \lim_{t \to 0}(f * \Phi_t)(a) = \lim_{t \to 0}\int \Phi_t(a - x)f(x)\,dx = f(a)$$

for almost every $a \in \mathbb{C}$. For an arbitrary distribution $f \in \mathcal{D}'(\Omega, \mathbb{V})$ we cannot speak of the integral at (5.5), but its convolution with Φ_t can nevertheless be formally defined. It is a function on the set $\Omega_t = \{a \in \Omega; \mathrm{dist}(a, \partial\Omega) > t\}$ given by

(5.6) $$f_t(a) = (f * \Phi_t)(a) = f[\Phi_t(a - \cdot)]$$

where we notice that the function $x \mapsto \Phi_t(a - x)$ belongs to $C_0^\infty(\Omega)$.

Which family of mollifiers we choose to fix is quite immaterial to our results. A good general reference here is the book by E. Stein [102].

Given any distribution $f \in \mathcal{D}'(\Omega, \mathbb{V})$ it is legitimate to write

(5.7) $$f_t(x) = f * \Phi_t(x)$$

for $x \in \Omega$, whenever $0 < t < \mathrm{dist}(x, \partial\Omega)$. Then we can define the mollified maximal function of f as

(5.8) $$(\mathcal{M}f)(x) = (\mathcal{M}_\Omega f)(x) = \sup\{|f_t(x)|;\ 0 < t < \mathrm{dist}(x, \partial\Omega)\}$$

for all $x \in \Omega$. Most often we shall ignore the subscript Ω, when the dependence of $\mathcal{M}$ on the domain need not be emphasized. For the Dirac delta, an easy computation shows that

$$(\mathcal{M}\delta)(x) = \frac{C}{|x|^2} \quad \text{for } x \in \mathbb{C} \setminus \{0\}.$$

5.2. Hardy-Orlicz Spaces

Now, the Hardy-Orlicz space $H^P(\Omega, \mathbb{V})$ is made up of Schwartz distributions $f \in \mathcal{D}'(\Omega, \mathbb{V})$ such that

(5.9) $$\|f\|_{H^P(\Omega)} = \|\mathcal{M}_\Omega f\|_{L^P(\Omega))} < \infty$$

Clearly, $H^P(\Omega, \mathbb{V})$ is a complete linear metric space with respect to the distance

(5.10) $$\mathrm{dist}(f, g) = \inf\left\{\frac{1}{\lambda} > 0;\ \lambda \int_\Omega P(\mathcal{M}(\lambda f - \lambda g)) \leq 1\right\}$$

If, moreover, P is convex the non-linear functional at (5.9) defines a norm, which makes $H^P(\Omega)$ a Banach space. For $\Omega = \mathbb{C}$, $\mathbb{V} = \mathbb{R}$ and $P(t) = t^p$ with $0 < p \leq 1$, our definition results in the classical Hardy spaces $H^p(\mathbb{C})$. Although it is not immediate from this definition, for sufficiently regular domains (for example Lipschitz domains) the Hardy space $H^1(\Omega)$ consists of the restrictions to Ω of functions in $H^1(\mathbb{C})$, [85, 29, 76]. Also note that $H^1(\Omega) \subset L^1(\Omega)$ and $\|f\|_{L^1(\Omega)} \leq \|f\|_{H^1(\Omega)}$.

In [65] the following lemma is established. A *regular distribution* is one which is absolutely continuous with respect to Lebesgue measure.

LEMMA 5.1. *Let P be any Orlicz function. A necessary and sufficient condition that all positive distributions in $H^P(\Omega)$ should be regular is:*

(5.11) $$\int_1^\infty \frac{P(s)ds}{s^2} = \infty$$

A stronger result is in fact true. To present this we associate to each P a new function

(5.12) $$R(t) = P(t) + t\int_0^t s^{-2}P(s)ds$$

where we may assume without loss of generality that $\int_0^1 s^{-2}P(s)ds < \infty$. Note that R grows faster than a linear function. Then, see [65],

THEOREM 5.2. *Under the condition (5.11) any positive distribution $f \in H^P(\Omega)$ is a function in $L^R_{loc}(\Omega)$. Furthermore, for each relatively compact subset $\Omega' \subset \Omega$ we have a uniform bound*

(5.13) $$\|f\|_{L^R(\Omega')} \leq C_P(\Omega') \|f\|_{H^P(\Omega)}$$

To some degree, the converse also holds; each function in $L^R_{loc}(\Omega)$ represents a distribution in $H^P_{loc}(\Omega)$, that is on relatively compact subdomains. When Ω is a cube we actually have the inclusion $L^R(\Omega) \subset L^P(\Omega)$ and the uniform bound

(5.14) $$\|f\|_{H^P(\Omega)} \leq C_P \|f\|_{L^R(\Omega)}$$

Formula (5.12) plays significant rôle in the theory of maximal inequalities. For example, when $P(t)$ grows linearly, $P(t) = t - \log(1+t) \approx t$ formula (5.12) gives $R(t) = t \log(1+t)$. As a particular case, we obtain the result of E. Stein [**103**].

THEOREM 5.3. *A non-negative function f belongs to $H^1_{loc}(\Omega)$ if and only if $f \log f \in L^1_{loc}(\Omega)$.*

5.3. BMO

Let us emphasize that the nature of a distribution in $H^P(\Omega)$ is determined not only by its size but also on its internal cancellation properties. These properties are perfectly visible in the atomic decompositions which we shall now turn to.

A measurable function $a(z)$ supported in some ball **B** is called an $\mathcal{H}^1$-*atom* if it satisfies both the conditions

(5.15) $$|a(z)| \leq \frac{1}{|\mathbf{B}|} \quad \text{a.e.} \quad z \in \mathbb{C}$$

(5.16) $$a_{\mathbf{B}} = \frac{1}{|\mathbf{B}|} \int_{\mathbf{B}} a(z) dz = 0$$

A function $f \in L^1(\mathbb{C})$ belongs to $\mathcal{H}^1(\mathbb{C})$ if and only if it can be written as a (possibly infinite) linear combination of $\mathcal{H}^1$-atoms, $f = \sum_{k=1}^{\infty} \lambda_k a_k$, with $\sum_{k=1}^{\infty} |\lambda_k| < \infty$. The norm is then defined by

(5.17) $$\|f\|_{\mathcal{H}^1} = \inf \left\{ \sum_{k=1}^{\infty} |\lambda_k| : f = \sum_{k=1}^{\infty} \lambda_k a_k \right\}$$

where the infimum is taken over all atomic decompositions of f. It is important to notice that such an f satisfies the *moment condition*

(5.18) $$\int_{\mathbb{C}} f(z) dz = 0.$$

Next, for a measurable function $g : \Omega \to \mathbb{V}$ and a ball $\mathbf{B} \subset \Omega$ we define the average of g on **B** as

(5.19) $$g_{\mathbf{B}} = \frac{1}{|\mathbf{B}|} \int_{\mathbf{B}} g(z) dz$$

If $g \in L^1_{loc}(\Omega, \mathbb{V})$ and if the norm

(5.20) $$\|g\|_{BMO} = \sup_{\mathbf{B}} \frac{1}{|\mathbf{B}|} \int_{\mathbf{B}} |g(z) - g_{\mathbf{B}}| dz < \infty,$$

then we say g is of *bounded mean oscillation*, $g \in BMO(\Omega, \mathbb{V})$.

There are two central facts to be noted here. The first is the duality theorem of Fefferman which states that $BMO(\mathbb{C})$ is the dual space of $\mathcal{H}^1(\mathbb{C})$, and also a result of Sarason which states that $\mathcal{H}^1(\mathbb{C})$ is the dual space of $VMO(\mathbb{C})$, this latter space being the completion of $C_0^\infty(\mathbb{C})$ in $BMO(\mathbb{C})$. In particular we note the following [**35**].

THEOREM 5.4. *There is a constant C such that if $f \in \mathcal{H}^1(\mathbb{C})$ and $g \in BMO(\mathbb{C})$, then*

$$\left| \int_\mathbb{C} f(x) g(z) dz \right| \leq C \|f\|_{\mathcal{H}^1} \|g\|_{BMO} \tag{5.21}$$

In general the integral (5.21) does not converge, however there are a number of ways to give meaning to it, see the recent [**25**] for a fairly complete discussion of this.

Finally the fundamental integrability properties of BMO–functions are found in the well–known John–Nirenberg lemma, [**67**]

THEOREM 5.5. *There exists a constant $\Theta > 0$ such that for every $h \in BMO(\Omega, \mathbb{V})$ and every ball or cube $\mathbf{B} \subset \mathbb{C}$, we have*

$$\frac{1}{|\mathbf{B}|} \int_\mathbf{B} \exp\left(\frac{\Theta |h(z) - h_\mathbf{B}|}{\|h\|_{BMO}} \right) dz \leq 2. \tag{5.22}$$

5.4. $L \log L$–Integrability

We will need to recall an important $L \log L$–estimate for the Jacobian. These estimates are carefully described in §7.3 [**58**].

THEOREM 5.6. *The Jacobian $J(z, h)$ of an orientation preserving mapping $h \in W^{1,2}(\Omega, \mathbb{C})$ is locally $L \log L$–integrable. More precisely, for concentric balls $\mathbf{B} \subset 2\mathbf{B} \subset \Omega$*

$$\int_\mathbf{B} J(z, h) \log\left(e + \frac{J(z, h)}{J_\mathbf{B}} \right) dz \leq C \int_{2\mathbf{B}} |Dh(z)|^2 \, dz \tag{5.23}$$

Here we have continued to use the notation $J_\mathbf{B}$ to denote the average of $J(z, h)$ over the ball $\mathbf{B}$. Also the term orientation preserving pertains to mappings with non-negative Jacobian.

This rather surprising higher integrability property of the Jacobian was first shown by S. Müller [**90**] by using local estimates similar to those of the $L \log L$–inequality for the maximal operator. We shall see that the explicit bound at (5.23) is a useful addition to Müller's result. It should be pointed out here that this theorem can be traced back to closely related earlier work of Gehring on the reverse Hölder inequalities.

In Chapter 8.3 we give related, but more general, results on the higher integrability of the Jacobian in Orlicz spaces that can be used to establish more refined versions of the results presented here, see Theorem 8.5.

5.5. Liouville Type Theorems

Here is a first taste of the power of Theorem 4.2. We use it to prove a Liouville type theorem for mappings of finite distortion. It will imply uniqueness of solutions to the Beltrami equation in certain circumstances, however we do not use it for this purpose here since we shall be able to prove a stronger result in due course.

5.5. LIOUVILLE TYPE THEOREMS

As we are going to deal with the Orlicz–Sobolev spaces $W^{1,P}(\Omega)$ in an unbounded domain, the behaviour of the defining function $P = P(t)$ for small values of t might be important. To avoid discussing rather delicate questions concerning the convergence of certain integrals near ∞ we assume henceforth that there are positive constants c_1 and c_2 such that

$$(5.24) \qquad c_1 t^2 \leq P(t) \leq c_2 t^2, \qquad 0 \leq t \leq 1.$$

Notice that in this case the Hölder conjugate function $Q = Q(t)$ also satisfies (5.24). We shall impose further conditions on P and Q for large values of t as the need arises.

THEOREM 5.7. *Let $f : \mathbb{C} \to \mathbb{C}$ be a mapping of finite distortion whose differential belongs to $L^P(\mathbb{C})$ with P an Orlicz function satisfying (4.26) and*

$$(5.25) \qquad P(t) = \frac{t^2}{L(t)}$$

with $\lim_{t \to \infty} L(t) = \infty$. Then f is constant.

In particular if $f \in W^{1,P}(\mathbb{C})$ with

$$P(t) = \frac{t^2}{\log(e+t) \log\log(3+t)}$$

then f is constant.

Proof. The Monotonicity Theorem 4.2 and the modulus of continuity estimate at Theorem 4.4 give us the bound

$$(5.26) \qquad |f(a) - f(b)| \leq CR \, \|Df\|_{\mathbf{B},P} \, \Xi\left(\frac{|a-b|}{2R}\right)$$

for $a, b \in \mathbf{B} = \mathbf{B}(R)$. Let us fix a and vary b with $|b| = R > |a|$ to find

$$(5.27) \qquad |f(a) - f(b)| \leq CR \, \|Df\|_{\mathbf{B},P}$$

for some constant C as Ξ is continuous and bounded on $[0, \tfrac{1}{2}]$. What we want to do now is to show that

$$(5.28) \qquad R \, \|Df\|_{\mathbf{B},P} \to 0$$

as $R \to \infty$.

LEMMA 5.8. *With P as above and $\varphi \in L^P(\mathbb{C})$ we have*

$$(5.29) \qquad R \, \|\varphi\|_{\mathbf{B},P} \to 0$$

where $\mathbf{B} = \mathbf{B}(R)$.

Proof. To this end we recall that from the definition of average and the form of P we have

$$\pi R^2 \|\varphi\|_{P,\mathbf{B}}^2 = \int_{\mathbf{B}} \frac{\varphi^2}{L\left(\frac{\varphi}{\|\varphi\|_{\mathbf{B},P}}\right)} \leq \int_{\mathbb{C}} \frac{\varphi^2}{L\left(\frac{\varphi}{\|\varphi\|_{\mathbf{B},P}}\right)}$$

Now it is clear from this that $\|\varphi\|_{\mathbf{B},P} \to 0$. Thus the Lebesgue dominated convergence theorem shows this last term tends to 0 and the lemma follows. □

The lemma and (5.27) now imply that f has a well defined limit at ∞ and this limit is $f(a)$. But since a was arbitrary, we conclude f is constant. □

It is to be remarked that the result clearly holds in much more generality. If one has two mappings of finite distortion f^1 and f^2 solving the same elliptic PDE and such that $D(f^1 - f^2) \in L^P(\mathbb{C})$, then of course the same proof shows $f^1 \equiv f^2$ modulo constants. Note in particular that we use the fact that in these circumstances the difference of mappings of finite distortion has finite distortion.

CHAPTER 6

The Principal Solution

We now present a fundamental concept of the theory of the planar Beltrami equation, namely that of the principal solution. Such a solution will define the domain of definition of all other solutions. We have discussed the history of this subject in our introduction. Here we aim to present some recent developments in a different light, in particular generalizations of David's existence theorem for the Beltrami equation when the distortion is assumed exponentially integrable. We try to place these results in a modern setting, using some of the powerful results available from harmonic analysis. The reader should note in passing that the proofs also establish holomorphic dependence on parameters, Corollary 7.16.

6.1. Solutions

Let $\mu = \mu(z)$ be a complex valued measurable function supported in a compact subset of $\mathbb{C}$ with $|\mu(z)| < 1$ almost everywhere.

Definition. We use the term *principal solution* to describe a homeomorphism $h : \overline{\mathbb{C}} \to \overline{\mathbb{C}}$ such that

(1) There is a discrete set E (the singular set) such that $h \in W^{1,1}_{loc}(\mathbb{C} \setminus E)$.
(2) The Beltrami equation

$$h_{\bar{z}}(z) = \mu(z) h_z(z)$$

holds for almost every $z \in \mathbb{C}$
(3) we have the normalization at ∞

$$h(z) = z + o(1)$$

The motivation for the last two parts of the definition are clear. To understand the first we note that some absolute continuity condition is really necessary if we are to consider a geometric theory of mappings where we might want to integrate partial derivatives, or use change of variables and so forth. If the complex partials ∂h and $\bar{\partial} h$ exist almost everywhere, it follows from a result of Gehring and Lehto [40] that h, being a homeomorphism, is in fact differentiable almost everywhere. However this still does not guarantee an ACL property. One might reasonably ask that $h \in W^{1,1}_{loc}$. As we shall see, however, there are natural circumstances where there is a nonempty singular set. Moreover, it will become clear that the key to understanding the Beltrami equation and its local solutions is in the existence and uniqueness properties of the principal solutions.

Let Ω be a domain in $\mathbb{C}$. We call any function f, not necessarily a homeomorphism, a *very weak solution* if it satisfies the two conditions:

- there is a discrete set $E_f \subset \Omega$ (the singular set) such that $h \in W^{1,1}_{loc}(\Omega \setminus E_f)$.

- the Beltrami equation

$$f_{\bar{z}}(z) = \mu(z) f_z(z)$$

holds for almost every $z \in \Omega$

The point here is that one expects a solution f to be simply the composition of the principal solution h with a meromorphic function φ defined on $h(\Omega)$,

$$f = \varphi \circ h,$$

(the Stoilow Factorization Theorem). Thus away from the poles, the solution f is locally as good as the principal solution. We call such a solution a *true* solution, whereas those very weak solutions which do not admit such a factorization are called *false solutions*.

We point out that the possibility of false solutions already occurs in the classical setting. In [57] we give an example, which we shall refine and extend in a later section.

6.2. Uniqueness of Principal Solutions

Here is the most general uniqueness result that we are aware of.

THEOREM 6.1. *Every elliptic equation*

$$h_{\bar{z}} = H(z, h_z)$$

admits at most one principal solution in the Sobolev-Orlicz class $z + W^{1,P}(\mathbb{C})$ whenever

$$P(t) \geq \frac{c\,t^2}{\log(e+t)\log\log(3+t)}$$

In the above and in what follows we use the term $z + W^{1,P}(\mathbb{C})$ to denote the class of mappings $h : \mathbb{C} \to \mathbb{C}$ such that $|h_{\bar{z}}| + |h_z - 1| \in L^P(\mathbb{C})$.

Proof. We first recall the hypotheses imply that there is a measurable compactly supported function $k : \mathbb{C} \to \mathbf{B}$ such that for each principal solution h

$$|H(z, \zeta) - H(z, \xi)| \leq k(z)|\zeta - \xi|$$

as $h_{\bar{z}}(z) = 0$ for z sufficiently large. The point of the proof is that given two principal solutions h^1 and h^2, the mapping $f = h^1 - h^2$ has finite distortion and its differential $Df = Dh^1 - Dh^2$ belongs to $L^P(\mathbb{C})$. To see this note

$$\begin{aligned}|(h^1 - h^2)_{\bar{z}}| &= |h^1_{\bar{z}} - h^2_{\bar{z}}| \\ &= |H(z, h^1_z) - H(z, h^2_z)| \\ &\leq k(z)|(h^1 - h^2)_z|\end{aligned}$$

whence $J(z, h^1 - h^2) \geq 0$. It then follows that f is constant from the Liouville type theorem. The normalization at ∞ immediately implies that this constant is 0. □

6.3. Stoilow Factorization

In this section we want to show that if a Beltrami equation admits a homeomorphic solution of Orlicz-Sobolev class $W^{1,P}_{loc}(\Omega)$, then all other solutions in the same class are obtained from this solution via composition with a holomorphic mapping. Again a key ingredient is monotonicity. In this section we make the standing assumption

$$P(t) \geq \frac{c\,t^2}{\log(e+t)\log\log(3+t)}$$

on the Orlicz function P.

However, the careful reader will see how the results presented in the appendices can be used to prove more delicate versions of the following theorems for Orlicz functions satisfying the divergence criteria (4.27).

6.3.1. Factorization Theorem. We begin with the following theorem

THEOREM 6.2. *Suppose we are given a homeomorphic solution $h \in W^{1,P}_{loc}(\Omega)$ to the Beltrami equation*

(6.1) $$h_{\bar{z}} = \mu(z) h_z \qquad a.e. \ \Omega$$

Then every very weak solution $f \in W^{1,P}_{loc}(\Omega)$ takes the form

$$f(z) = \phi(h(z)), \qquad z \in \Omega$$

where $\phi : h(\Omega) \to \mathbb{C}$ is holomorphic.

Proof. First observe that (6.1) implies the map f has finite distortion. By Theorem 4.5 f is continuous. Fix an arbitrary closed disk $\overline{\mathbf{B}} \subset h(\Omega)$ and set $\varphi = f \circ h^{-1} : \partial \mathbf{B} \to \mathbb{C}$. Then φ is a continuous function. The solution to the classical Dirichlet problem tells us that there is a function u continuous on $\overline{\mathbf{B}}$ which is harmonic in $\mathbf{B}$ and equal to $\Re(\varphi)$ on $\partial \mathbf{B}$. Let $v \in C^\infty(\mathbf{B})$ be the harmonic conjugate to u on $\mathbf{B}$. Then the function $\phi = u + iv$ is holomorphic on $\mathbf{B}$ (the reader is cautioned about the continuity of ϕ on $\overline{\mathbf{B}}$). Consider the two functions $\phi \circ h$ and f defined and continuous on $h^{-1}(\overline{\mathbf{B}})$. By construction, these two functions have the same real part on $\partial h^{-1}(\overline{\mathbf{B}})$. Now both of these functions lie in $W^{1,P}_{loc}(h^{-1}(\mathbf{B}))$ and satisfy the same Beltrami equation. Therefore their difference, $g = f - \phi \circ h$ is a mapping of finite distortion. Theorem 4.2 shows that the real part of g is weakly monotone, continuous in $h^{-1}(\overline{\mathbf{B}})$, and vanishes on the boundary of this set. We deduce from Lemma 4.1 that $\Re(g) \equiv 0$ in $h^{-1}(\mathbf{B})$. In particular we see that $J(z,g) \equiv 0$ in $h^{-1}(\mathbf{B})$ and as g is a mapping of finite distortion, $Dg(z) \equiv 0$ in $h^{-1}(\mathbf{B})$. From this we deduce that $g(z) \equiv c$ an imaginary constant. We therefore have the factorization

$$f(z) = \phi(h(z)) + c, \qquad z \in \mathbf{B}$$

Recall that $\mathbf{B}$ was an arbitrary disk compactly contained in $h(\Omega)$. As f is continuous in Ω the principle of analytic continuation gives us a unique extension of $\phi + c$ to a holomorphic function, call it ϕ again, defined in $h(\Omega)$ and such that $f = \phi \circ h$. □

The following refinement of this theorem follows directly from the removability of singularities for holomorphic functions.

COROLLARY 6.3. *Let $\mu : \Omega \to \mathbb{C}$ be a Beltrami coefficient, not necessarily compactly supported. Suppose $h : \Omega \to \mathbb{C}$ is a homeomorphism of class $W^{1,P}_{loc}(\Omega \setminus E)$, with E discrete, and suppose h solves the Beltrami equation*

(6.2) $$h_{\bar{z}} = \mu(z) h_z \quad \text{a.e.} \quad \Omega$$

Suppose that $f \in W^{1,P}_{loc}(\Omega \setminus E')$, E' discrete, is another solution which is continuous in Ω. Then there is a holomorphic map $\phi : g(\Omega) \to \mathbb{C}$ such that $f = \phi \circ h$.

Note that any solution of the form $\Psi \circ h$ has the regularity required of f (i.e. the same as h) and so such an assumption is necessary. The factorization theorem shows that a homeomorphic solution generates all solutions that are as regular as itself.

Also note that the factorization theorem implies that if an equation admits a homeomorphic solution $h : \Omega \to \mathbb{C}$ in the class $W^{1,P}_{loc}(\Omega \setminus E)$, then all continuous solutions $f : \Omega \to \mathbb{C}$ in $W^{1,P}_{loc}(\Omega \setminus E')$ are open and discrete.

6.3.2. Failure of Factorization. In this section we present an example worked out with generous help from Kari Astala. The example shows that even for fairly nice solutions one cannot expect a general factorization theorem even in the case of bounded distortion.

THEOREM 6.4. *Let $K > 1$ and $q_0 < \frac{2K}{K+1}$. Then there is a Beltrami coefficient μ supported in the unit disk with the following properties.*

- $\|\mu\|_\infty = \frac{K-1}{K+1}$,
- *The Beltrami equation $h_{\bar{z}} = \mu h_z$ admits a Hölder continuous solution $h \in z + W^{1,q_0}(\mathbb{C})$ which fails to be in $W^{1,2}_{loc}(\mathbb{C})$,*
- *The solution h is not quasiregular, and therefore not the principal solution, nor obtained from the principal solution by factorization.*

Proof. Choose a regular Cantor set E in the unit disk $\mathbf{B}$ of Hausdorff dimension $1 + \epsilon$, where $0 < \epsilon < 1$ is to be determined. Let ν denote the $(1 + \epsilon)$-Hausdorff measure defined on compact subsets of $\mathbb{C}$. We set

$$\varphi(w) = \frac{1}{2\pi i} \int_E \frac{d\nu(\zeta)}{w - \zeta}$$

as the Cauchy transform of ν. Then φ is a Hölder continuous (with exponent ϵ) function defined in $\mathbb{C}$ which is holomorphic in $\mathbb{C} \setminus E$ and $\varphi(w) = \frac{C}{w} + o(1)$ as $w \to \infty$.

We have the estimate

$$|\varphi'(w)| \leq \frac{c}{\text{dist}(w, E)^{1-\epsilon}}$$

for some positive constant c, see [36]. The function φ is certainly not in the appropriate Sobolev space as it is not holomorphic. We are going to improve the regularity of φ at the cost of adding some distortion.

To this end we find a K-quasiconformal mapping $g : \mathbb{C} \to \mathbb{C}$ with $g(z) = z$ outside the unit disk, $g|\mathbb{C} \setminus F \to \mathbb{C} \setminus E$ smooth, and such that $g(F) = E$ where F another regular Cantor set of smaller Hausdorff dimension $1+\alpha$, where $-1 < \alpha < \epsilon$. The distortion and dimension can be shown to be related by the formula,

(6.3) $$K = \frac{1-\alpha}{1+\alpha} \frac{1-\epsilon}{1+\epsilon} > 1,$$

6.3. STOILOW FACTORIZATION

see [6]. Rearranging, gives
$$\alpha = \frac{1 - \frac{1-\epsilon}{1+\epsilon} K}{1 + \frac{1-\epsilon}{1+\epsilon} K}$$

The construction of such Cantor sets and such a mapping g, while not obvious, follow from standard results in the theory of quasiconformal mappings.

The mapping g also comes with the estimates
$$\text{dist}(g(z), E) \approx \text{dist}(z, F)^{1/K}, \qquad |g'(z)| \approx \text{dist}(z, F)^{-1+1/K}$$

We now set
$$f(z) = (\varphi \circ g)(z)$$

Then $f(z) = \frac{1}{z} + o(1)$ near ∞. Now g is quasiconformal and so in $W_{loc}^{1,2}(\mathbb{C})$. The composition of the two mappings f and g is readily seen to be ACL. Therefore we need only check the degree of integrability of the differential of f. Pointwise, we have

(6.4) $$|Df(z)| = |\varphi'(g(z))| \, |Dg(z)|$$

so near the set F we have
$$|Df(z)| \leq \left(\frac{c_1}{\text{dist}(z,F)}\right)^{(1-\epsilon)/K + 1 - 1/K} = \left(\frac{c_1}{\text{dist}(z,F)}\right)^{1-\epsilon/K}$$

Now as F is a regular Cantor set $\text{dist}^{-1}(z, F)$ is integrable with any power smaller than $2 - \dim(F) = 1 - \alpha$. We solve
$$q_\epsilon(1 - \epsilon/K) = (1 - \alpha)$$

to obtain

(6.5) $$q_\epsilon = \frac{2K}{(K + \frac{1+\epsilon}{1-\epsilon})(1 - \frac{\epsilon}{K})}$$

Then we have $f \in W_{loc}^{1,q}$ for all $q < q_\epsilon$.

We now have to adjust ϵ so that $q_\epsilon > q$. This is possible since $q_\epsilon \to \frac{2K}{K+1}$ as $\epsilon \to 0$.

We next have to invert this mapping to get the normalization at ∞ that we wanted. Finally the classical regularity theory discussed in the introduction shows this result to be sharp. □

CHAPTER 7

Solutions for Integrable Distortion

We have seen in the previous chapter how the homeomorphic solutions generate all local solutions in the same class. Here we give a condition for the existence of a homeomorphic solution under the supposition that there is a fairly regular nonconstant solution. Much of this is known, and some of it follows from the above so we do not give complete proofs.

Let us make a few normalizations to structure our discussion in this chapter.

(1) μ is compactly supported in the unit ball $\mathbf{B}$.
(2) The distortion function $K(z) = \frac{1+|\mu(z)|}{1-|\mu(z)|}$ lies in $L^1(\mathbf{B})$.
(3) The Beltrami equation admits a nonconstant solution $f \in W^{1,2}(\mathbf{B})$, that is

$$f_{\bar{z}} = \mu f_z, \qquad \text{a.e. } \mathbf{B}$$

We wish to stress here that the planar theory of non-linear elasticity deals with mappings $f \in W^{1,2}(\mathbf{B})$ with integrable distortion. Such mappings are usually obtained as minimizers of certain energy integrals. Thus in fairly natural circumstances the existence of at least one $W^{1,2}(\mathbf{B})$ solution is possible. It is in general, however, quite a strong assumption. (However, one should jump ahead for a moment to consider this result in the light of Theorem 7.2).

We now present the following theorem.

THEOREM 7.1. *Under the hypotheses (1), (2) and (3) above, there exists a principal solution to the equation*

(7.1) $$h_{\bar{z}}(z) = \mu(z) h_z(z), \qquad \text{a.e. } \mathbb{C}$$

which has the following additional properties:

- *Partial Regularity; there is a finite set E such that $h \in W^{1,2}_{loc}(\mathbb{C} \setminus E)$.*
- *Factorization; each solution $g \in W^{1,2}_{loc}(\Omega)$ to the equation*

$$g_{\bar{z}}(z) = \mu(z) g_z(z), \qquad \text{a.e. } \Omega$$

admits a Stoilow factorization

(7.2) $$g(z) = (\Phi \circ h)(z)$$

where Φ is holomorphic in $h(\Omega)$. In particular, all non-constant solutions in $W^{1,2}_{loc}(\Omega)$ are open and discrete. Here Ω is an arbitrary domain in $\mathbb{C}$.
- *Uniqueness; Let E' be a finite set and $h_0 \in W^{1,2}_{loc}(\mathbb{C} \setminus E')$ be a solution to the Beltrami equation, continuous in $\mathbb{C}$, with $h_0(z) = z + o(1)$ near ∞. Then $h_0 = h$, the principal solution.*

- **Inverse;** Denote the inverse mapping of h by $g = g(w)$. Then g has no singularities; $g \in w + W^{1,2}(\mathbb{C})$ with the precise estimate

$$\int_{\mathbb{C}} (|g_{\bar{w}}|^2 + |g_w - 1|^2) dw \leq C \int_{\mathbf{B}} K(z) dz$$

- **Modulus of continuity;** we have the modulus of continuity estimate

$$|g(a) - g(b)|^2 \leq C \frac{1 + |a|^2 + |b|^2}{\log\left(1 + \frac{|a|+|b|}{|a-b|}\right)} \int_{\mathbf{B}} K(z) dz.$$

In both these last two estimates C is an absolute constant.

Sketch of proof. This result is basically a refinement of the factorization theorem in [64] and we shall rely on that proof here, see too [58]. One begins by considering the truncated Beltrami equation

$$h_{\bar{z}}^\epsilon(z) = \mu_\epsilon(z) h_z^\epsilon(z), \qquad |\mu_\epsilon| < 1 - \epsilon$$

where

$$\mu_\epsilon(z) = \begin{cases} \mu(z) & \text{if } |\mu(z)| \leq 1 - \epsilon \\ (1-\epsilon)\mu(z)/|\mu(z)| & \text{otherwise} \end{cases}$$

and with its principal solution $h^\epsilon \in z + W^{1,2}(\mathbb{C})$. When $\epsilon \to 0$ it was shown that the inverse mappings $g_\epsilon = h_\epsilon^{-1}$ converge locally uniformly in $\mathbb{C}$ to a mapping g. More importantly it turns out that g is a homeomorphism on $g^{-1}(\mathbf{B})$, where we emphasize that $\mathbf{B}$ is the domain of the nonconstant solution in $W^{1,2}(\mathbf{B})$ to the original equation. The mapping g lies in the space $W^{1,2}(g^{-1}(\mathbf{B}))$ and we have the estimates

$$\int_{g^{-1}(\mathbf{B})} |Dg(w)|^2 \, dw \leq \int_{\mathbf{B}} K(z, f) \, dz$$

and

$$|g(a) - g(b)|^2 \leq \frac{C(R^2 + \int_{\mathbf{B}} K(z,f) \, dz)}{\log \frac{2}{|a-b|}}$$

whenever $a, b \in B(0, R)$, $R \geq 1$ and $|a - b| < 2$ and C is an absolute constant.

We denote the inverse to g by $h : \mathbf{B} \to g^{-1}(\mathbf{B})$. It was further shown that every $W_{loc}^{1,2}(\Omega)$ solution to the Beltrami equation admits a factorization, so that in fact

$$g(z) = \Phi(h(z))$$

We now use this to prove that g is a homeomorphism of $\overline{\mathbb{C}}$. To see this, note that g was obtained as a local uniform limit of homeomorphisms $g_\epsilon : \overline{\mathbb{C}} \to \overline{\mathbb{C}}$ which are conformal in $g_\epsilon^{-1}(\mathbb{C} \setminus F)$ where F is a relatively compact subset of $\mathbf{B}$, namely the support of μ. Each has a Taylor expansion of the form

$$g_\epsilon(w) = w + \frac{a_1^\epsilon}{w} + \frac{a_2^\epsilon}{w^2} \cdots$$

The family $\{g_\epsilon\}_{\epsilon>0}$ of conformal maps must have a nonconstant conformal map as a limit as the limit has already shown to be nonconstant on $\mathbf{B} \setminus F$. In this way g extends to $\overline{\mathbb{C}}$ homeomorphically. Now we may extend h by setting $h = g^{-1} : \overline{\mathbb{C}} \to \overline{\mathbb{C}}$. This is the principal solution we seek. The only things worthy of any further explanation are the partial regularity and uniqueness.

Clearly h is holomorphic in $\mathbb{C} \setminus F \supset \mathbb{C} \setminus \mathbf{B}$. Concerning the regularity in $\mathbf{B}$ we use the factorization
$$f(z) = \Phi(h(z))$$
where f is our given nonconstant solution in $W^{1,2}_{loc}(\mathbf{B})$. Φ is holomorphic and nonconstant in $h^{-1}(\mathbf{B})$ and therefore has only a finite number of critical points on any relatively compact subdomain. Away from these critical points Φ has a holomorphic inverse and the partial regularity of h follows.

Concerning uniqueness, if $h_0 \in W^{1,2}_{loc}(\mathbb{C} \setminus E')$ is another principal solution, continuous in $\mathbb{C}$, we may write
$$h_0(z) = \Phi(h(z))$$
where Φ is holomorphic in $h^{-1}(\mathbb{C} \setminus (E \cap E'))$. The point here being that the singular set $h^{-1}(E \cap E')$ is countable (having finite linear measure would suffice) and therefore removable for the locally bounded Φ. Since Φ is now linear and near ∞ we have $\Phi(w) = w + o(1)$ it follows that $h_0 = h$. $\square$

7.1. Distortion in the Exponential Class

It is plain that establishing the existence of a principal solution, without assuming the existence of a local solution, requires a higher degree of regularity of the distortion function. As we have repeatedly stressed, and as is again evident from the previous section, the Sobolev space $W^{1,2}_{loc}(\mathbb{C})$ is the most natural space to go looking for a solution to the Beltrami equation. The reader should note that it is in this class that the Jacobian determinant $J(z,h) = |h_z|^2 - |h_{\bar{z}}|^2$ is locally integrable, a property one cannot dispense with. On the other hand it is known that Jacobians of orientation preserving mappings (that is with $J(z,f) \geq 0$) are locally integrable under slightly weaker assumptions than $W^{1,2}_{loc}$. This is discussed in some detail in [58].

Recall that the distortion functions $K(z)$ of a solution to the Beltrami equation is defined by

(7.3) $$K(z) = \frac{1 + |\mu(z)|}{1 - |\mu(z)|}$$

and therefore $1 \leq K(z) < \infty$ almost everywhere, and $K(z) = 1$ outside the support of μ.

What we are about to examine now is an explicit example close in spirit to those of §3 however this time we will have continuity. This example serves as a guide for the results one might obtain. But first let us provide some motivation. It is clear that one cannot stray too far from the case of bounded distortion. The first natural condition that comes to mind is the assumption

(7.4) $$e^K \in L^p_{loc}(\mathbb{C})$$

for some positive p. The line of reasoning is as follows. The Beltrami equation yields the distortion inequality

(7.5) $$|Dh(z)|^2 \leq K(z) J(z,h)$$

Now if for some reason the Jacobian function $J = J(z,h)$ is locally integrable (for instance if $h \in W^{1,1}_{loc}(\mathbb{C})$ is a homeomorphism), we may use the elementary inequality

(7.6) $$ab \leq a \log(1 + a) + e^b - 1$$

together with the facts that $t \mapsto t/\log(e+t)$ is an increasing function and that pointwise $|Dh|^2 \leq KJ$ to find that

$$\begin{aligned}
\frac{|Dh|^2}{\log(e+|Dh|^2)} &\leq \frac{KJ}{\log(e+KJ)} \\
&\leq \frac{KJ}{\log(e+J)} \\
&\leq \frac{J}{p\log(e+J)} \log\left(1 + \frac{J}{p\log(e+J)}\right) + e^{pK} - 1 \\
&\leq \frac{2J}{p} \log\left(e + \frac{1}{p}\right) + e^{pK} - 1
\end{aligned}$$

for all $p > 0$. This last inequality follows as

$$\begin{aligned}
\log\left(1 + \frac{J}{p\log(e+J)}\right) &\leq \log\left(\left(1 + \frac{J}{\log(e+J)}\right)\left(1 + \frac{1}{p}\right)\right) \\
&= \log(e+J) + \log\left(e + \frac{1}{p}\right) \\
&\leq 2\log(e+J) \log\left(e + \frac{1}{p}\right)
\end{aligned}$$

as $x + y \leq 2xy$ if $x, y \geq 1$.

We now integrate the previous estimate to obtain

$$(7.7) \qquad \int_\Omega \frac{|Dh|^2}{\log(e+|Dh|)} \leq \frac{4}{p} \log\left(e + \frac{1}{p}\right) \int_\Omega J(z,h) + 2 \int_\Omega [e^{pK(z)} - 1]\, dz$$

for any bounded domain. This shows that h belongs to the Sobolev class $W^{1,Q}_{loc}(\mathbb{C})$ with

$$(7.8) \qquad Q(t) = \frac{t^2}{\log(e+t)}$$

Conversely, the $L^2 \log^{-1} L$–integrability of the differential of an orientation preserving mapping yields $L\log\log L$ integrability of the Jacobian, that is slightly better than L^1. This slight gain in the regularity of the Jacobian determinant is precisely why it is possible to study solutions to the Beltrami equation using the space $W^{1,Q}_{loc}$ as a starting point.

7.2. An Example

We believe the following example provides the precise limits of what is possible. Given a positive number θ we set

$$(7.9) \qquad h(z) = h_\theta(z) = \begin{cases} \frac{z}{|z|}\left(1 + \frac{1}{\theta}\log\frac{1}{|z|}\right)^{-\theta} & |z| \leq 1 \\ z & |z| > 1 \end{cases}$$

Following the computation in §2.3 we find the Beltrami coefficient of h_θ to be

$$(7.10) \qquad \mu = \mu_\theta(z) = \chi_{\mathbf{B}} \frac{z}{\bar{z}} \frac{\log|z|}{2\theta - \log|z|}$$

where, as usual, $\chi_{\mathbf{B}}$ denotes the characteristic function of the unit disk. This shows h to be a principal solution to the Beltrami equation

$$(7.11) \qquad \bar{\partial} h_\theta = \mu_\theta\, \partial h_\theta$$

The distortion function is finite except at the origin. We have

(7.12) $$K = K_\theta(z) = 1 + (\frac{1}{\theta}\log\frac{1}{|z|})\chi_\mathbf{B}$$

and the norm of the differential is given by

(7.13) $$|Dh_\theta(z)| = \begin{cases} \frac{1}{|z|}(1+\frac{1}{\theta}\log\frac{1}{|z|})^{-\theta} & |z|\le 1 \\ 1 & |z|>1 \end{cases}$$

From this we see

The function $e^{K(z)} \in L^p_{loc}(\mathbb{C})$ if and only if $p < 2\theta$

Indeed,

(7.14) $$\int_\mathbf{B} e^{pK(z)}dz = \frac{2\theta e^p \pi}{2\theta - p}$$

On the other hand we find h_θ belongs to the natural Sobolev space; that is

$$h_\theta \in z + W^{1,2}(\mathbb{C})$$

if and only if $1 < 2\theta$. Indeed, we have

(7.15) $$\int_\mathbf{B} |Dh(z)|^2 dz = \frac{2\theta\pi}{1-2\theta}$$

This leads us to the following observation

In order that there should exist a principal solution in the natural Sobolev space $z + W^{1,2}(\mathbb{C})$ it is necessary that the exponent p at (7.4) is large, at least $p \ge 1$.

7.3. Results

Our first task is to validate the above observation. And the result we prove is in some sense a preliminary. It is an important refinement of a result due to David [33], whose methods were substantially different.

THEOREM 7.2. (Existence) *There exists a number $p_0 > 1$ such that every Beltrami equation*

$$h_{\bar{z}}(z) = \mu(z)\, h_z(z) \qquad a.e. \ \ \mathbb{C}$$

with Beltrami coefficient μ satisfying

$$|\mu(z)| \le \frac{K(z)-1}{K(z)+1}\chi_\mathbf{B}$$

and

(7.16) $$e^K \in L^p(\mathbf{B})$$

with $p \ge p_0$, admits a unique principal solution $h \in z + W^{1,2}(\mathbb{C})$.

It appears that the infimum of such numbers p_0, called the *critical exponent* for $W^{1,2}$–regularity, might be equal to 1.

While the proof of this theorem is quite involved, we have already seen the uniqueness established. The basic properties of the principal solution are listed in Theorem 7.1 which now applies. However we have the additional modulus of continuity estimate

$$|h(a) - h(b)|^2 \leq \frac{C_K}{\log(e + \frac{1}{|a-b|})} \tag{7.17}$$

for all $a, b \in 2\mathbf{B}$. An explicit bound for the constant C_K is given by

$$C_K \leq C \log^2 \left(\int_{\mathbf{B}} e^{pK} \right) \tag{7.18}$$

As a matter of fact, somewhat more is true. The higher the exponent of integrability of e^K the better the regularity of the solution. That is even beyond L^2. We also see a jump in regularity. Precisely we have

THEOREM 7.3. *For each $\alpha > 0$ there is a critical exponent $p_\alpha \geq 1 + \alpha$ such that if*

$$e^K \in L^p(\mathbf{B}), \qquad \text{for } p \geq p_\alpha$$

then there is a unique principal solution in $z + W^{1,P_\alpha}(\mathbb{C})$ with Orlicz function

$$P_\alpha(t) = t^2 \log^\alpha(e+t) \tag{7.19}$$

Moreover, every solution $f \in W^{1,Q_\alpha}_{loc}(\Omega)$,

$$Q_\alpha(t) = \frac{t^2}{\log^\alpha(e+t)}$$

to the Beltrami equation actually belongs to the space $W^{1,P_\alpha}_{loc}(\Omega)$.

Notice that (Q_α, P_α) is a Hölder conjugate pair of Orlicz functions, as in the classical case mentioned in §2.4.

The situation is rather different if the integrability exponent of e^K is smaller than the critical exponent p_0. Here we see from our example that the principal solution need not be in $z + W^{1,2}(\mathbb{C})$. The example puts the principal solution in $z + W^{1,Q_1}(\mathbb{C})$, and this is not by accident:

THEOREM 7.4. *Suppose the distortion function $K = K(z)$ for the Beltrami equation is such that $e^K \in L^p(\mathbf{B})$ for some positive p. Then the equation admits a unique principal solution h*

$$h \in z + W^{1,Q}(\mathbb{C}), \qquad Q(t) = t^2 \log^{-1}(e+t) \tag{7.20}$$

Moreover, every $W^{1,Q}_{loc}(\Omega)$ solution is a true solution.

In this class the factorization theorem is valid and we see that the principal solution generates all solutions in this class. Again a little more is true, however, this time it is a little more surprising. The $L^2 \log^{-1} L$-integrability of $|Dh|$ actually holds under much weaker assumptions on K. Namely if

$$\exp\left(\frac{K}{1 + \log K}\right) \in L^p(\mathbf{B})$$

provided p is sufficiently large. This is the topic of our next section.

7.4. Distortion in the Subexponential Class

In this section we present our main results regarding solutions to the Beltrami equation when the distortion function $K = K(z)$ satisfies

$$\exp\left(\frac{K}{1+\log K}\right) \in L^p(\mathbf{B}) \tag{7.21}$$

for some $p > 0$. We call such distortion functions *subexponentially* integrable. We recall our assumption here that the Beltrami coefficient is compactly supported in $\mathbf{B}$, and thus $K(z) \equiv 1$, $|z| \geq 1$. Again, we present an example to proscribe the limits of what one might obtain.

7.5. An Example

Given a positive number θ we set

$$h(z) = h_\theta(z) = \begin{cases} \frac{z}{|z|}\left(1 + \frac{1}{\theta}\log\log\frac{e}{|z|}\right)^{-\theta} & |z| \leq 1 \\ z & |z| > 1 \end{cases} \tag{7.22}$$

Much as before

$$\mu = \mu_\theta(z) = \chi_\mathbf{B}\frac{z}{\bar{z}}\frac{1-(1+\frac{1}{\theta}\log\log\frac{e}{|z|})\log\frac{e}{|z|}}{1+(1+\frac{1}{\theta}\log\log\frac{e}{|z|})\log\frac{e}{|z|}} \tag{7.23}$$

and

$$K = K_\theta(z) = 1 + \left(\frac{1}{\theta}\log\log\frac{e}{|z|}\right)\log\frac{e}{|z|}, \quad |z| \leq 1. \tag{7.24}$$

with the norm of the differential given by

$$|Dh_\theta(z)| = \begin{cases} \frac{1}{|z|}\left(1+\frac{1}{\theta}\log\log\frac{e}{|z|}\right)^{-\theta} & |z| \leq 1 \\ 1 & |z| > 1 \end{cases} \tag{7.25}$$

Hence

$$\frac{K(z)}{1+\log K(z)} = [\frac{1}{\theta} + o(1)]\log\frac{e}{|z|} \tag{7.26}$$

near 0 and so

the function $\exp\left(\frac{K}{1+\log K}\right) \in L^p_{loc}(\mathbb{C})$ if and only if $p < 2\theta$

On the other hand we have for all $\theta > 0$

$$\int_\mathbf{B} \frac{|Dh|^2\, dz}{\log(e+|Dh|)\log\log(3+|Dh|)} < \infty \tag{7.27}$$

However we have higher integrability of Dh if θ is large. Precisely

$$\int_\mathbf{B} \frac{|Dh|^2\, dz}{\log(e+|Dh|)} < \infty \tag{7.28}$$

if and only if $2\theta > 1$. We summarize as

CONJECTURE 7.5. *Suppose the distortion function K of a Beltrami equation satisfies (7.21) for $p > 1$. Then there is a unique principal solution in Sobolev space $z + W^{1,Q}(\mathbb{C})$, with $Q(t) = t^2 \log^{-1}(e+t)$. Consequently, every nonconstant solution in $W^{1,Q}_{loc}(\Omega)$ is open and discrete.*

7.5.1. Results.
However, what we shall prove is the following.

THEOREM 7.6. *There is a number $p_* \geq 1$ such that every Beltrami equation whose distortion function has*

$$\exp\left(\frac{K(z)}{1 + \log K(z)}\right) \in L^p(\mathbf{B}), \quad \mu(z) \equiv 0 \quad \text{outside } \mathbf{B},$$

for $p \geq p_*$, admits a unique principal solution $h \in z + W^{1,Q}(\mathbb{C})$ with Orlicz function $Q(t) = t^2 \log^{-1}(e+t)$. Moreover we have

- Modulus of Continuity;

(7.29) $$|h(a) - h(b)|^2 \leq \frac{C_K}{\log\log(1 + \frac{1}{|a-b|})}$$

for all $a, b \in 2\mathbf{B}$.

- Inverse; *The inverse map $g = h^{-1}(w)$ has finite distortion $\mathcal{K} = \mathcal{K}(w)$ and*

$$\log \mathcal{K} \in L^1(\mathbb{C})$$

- Factorization; *each solution $g \in W_{loc}^{1,Q}(\Omega)$ to the equation*

$$g_{\bar{z}} = \mu g_z, \quad \text{a.e. } \Omega$$

admits a Stoilow factorization

(7.30) $$g(z) = (\Phi \circ h)(z)$$

where Φ is holomorphic in $h(\Omega)$. In particular, all non-constant solutions in $W_{loc}^{1,Q}(\Omega)$ are open and discrete.

The uniqueness, modulus of continuity estimate and the factorization already follow from above, see Theorems 5.3 and 9.1. The hard part is existence. The proof of the properties of the inverse map are not too difficult and so we sketch them now. Since $\mathcal{K}(w) = K(z, h)$ we see $\mathcal{K}(w) = 1$ for all sufficiently large values of w. We need only show $\mathcal{K} \in L^1(h(\mathbf{B}))$. Integration by substitution yields

(7.31) $$\int_{h(\mathbf{B})} \log \mathcal{K}(w) \, dw = \int_{\mathbf{B}} \log K(z,h) J(z,h) \, dz$$

Regarding this change of variables (7.31) we point out that the principal solution, as we have constructed it, will be the limit of a sequence of quasiconfomal mappings with suitable uniform bounds, see §13.5. The change of variables can therefore be performed at the level of that sequence of mappings, following which a standard limiting argument legitimizes this procedure.

We next apply the elementary pointwise inequality

(7.32) $$J \log K \leq C_p J \log \log(e + J) + C_p \exp\left(\frac{pK}{1 + \log K}\right)$$

valid for all $J \geq 0$ and $K \geq 1$. Thus

$$\int_{h(\mathbf{B})} \log \mathcal{K} \leq C_p \int_{\mathbf{B}} J \log \log(e+J) + C_p \int_{\mathbf{B}} \exp\left(\frac{pK}{1 + \log K}\right)$$

This latter integral is finite by assumption. The crucial observation here is that the Jacobian determinant of the principal solution h belongs to $L \log \log L(\mathbf{B})$. this is in fact a general fact about orientation preserving mappings in $W_{loc}^{1,P}(\mathbb{C})$. There is now an extensive literature on this subject and we encourage the reader to look at [58, 20, 30, 43, 44, 63, 65, 50, 89]. In particular we have $\log \mathcal{K} \in L^1(\mathbb{C})$.

7.6. Further Generalities

In fact let us consider this inverse map for a moment longer. The inverse of the example at (7.22) can be computed to be

$$z = h^{-1}(w) = \frac{w}{\bar{w}} \exp\left(1 - e^{\theta(|w|^{-1/\theta} - 1)}\right) \qquad (7.33)$$

so we find that $\log \mathcal{K}(w)$ has dominant term $\theta |w|^{-1/\theta}$ and

$$\frac{\theta}{|w|^{1/\theta}} \in L^1(\mathbf{B}) \qquad (7.34)$$

if and only if $2\theta > 1$. This is the same condition that guarantees the L^1–integrability of $\exp(K/(1+\log K))$.

In much the same way as we lost $W^{1,2}$–regularity of the principal solution if the exponent p was smaller than the critical exponent, we should expect to loose regularity here. Though as before we have

THEOREM 7.7. *A Beltrami equation whose distortion function satisfies*

$$\exp\left(\frac{K(z)}{1 + \log K(z)}\right) \in L^p(\mathbf{B}) \qquad (7.35)$$

for some positive p has a unique principal solution

$$h \in z + W^{1,R}(\mathbb{C})$$

with

$$R(t) = \frac{t^2}{\log(e+t) \log \log(3+t)} \qquad (7.36)$$

Again of course, the factorization theorem and a modulus of continuity estimate are available.

We presume these sorts of results continue through a spectrum of Orlicz functions under weaker and weaker assumptions on the integrability of the distortion function. We have already shown that the limit of such extensions is given by the example of §3. One might reasonably ask the following.

Let $\Theta : [1, \infty] \to [1, \infty]$, be an increasing function with

$$\int_1^\infty \frac{\Theta(s)}{s^2} ds = \infty$$

Then is it true that every Beltrami equation whose distortion function satisfies

$$\int_{\mathbf{B}} \exp[\lambda \Theta(K)] < \infty, \quad \mu(z) \equiv 0 \text{ outside } \mathbf{B},$$

for some $\lambda > 0$ admits a unique principal solution $h \in z + W^{1,R}(\mathbb{C})$, with

$$R(t) = \frac{\Theta(t)}{1 + \int_1^t \frac{\Theta(s)}{s^2} ds} \quad ?$$

We believe the answer to this question is yes, with the proviso that some other minor condition on Θ may be necessary. The reader might care to consider the results presented in our appendices which justify our belief in an affirmative answer.

7.7. Existence Theory

We recall the Beltrami equation

$$f_{\bar{z}} = \mu(z) f_z \tag{7.37}$$

Typically what one assumes to prove existence and uniqueness are ellipticity bounds on the Beltrami coefficient μ, say $\|\mu\|_\infty < 1$. We will deal only with orientation preserving solutions, $J(z,f) = |f_z|^2 - |f_{\bar{z}}|^2 \geq 0$, thus $|\mu| < 1$ almost everywhere. The existence problem is to determine conditions on μ which guarantee a unique normalized solution (exactly what normalization will be discussed in a moment). These conditions take the form of integrability conditions on the distortion function

$$K(z) = \frac{1 + |\mu(z)|}{1 - |\mu(z)|} \tag{7.38}$$

Various obvious reductions show the important case to be when the Beltrami coefficient μ is compactly supported in the unit disk **B**. Then of course any solution is analytic outside the unit disk. This leads us to our normalization: a solution $f : \mathbb{C} \to \mathbb{C}$ to the Beltrami equation (7.37) is called a *principal solution* if it has the Taylor expansion at infinity

$$f(z) = z + \frac{a_1}{z} + \frac{a_2}{z^2} + \cdots$$

In which case f will take the form

$$z + T\omega(z) \tag{7.39}$$

where T is the complex Riesz Potential, defined at (2.16) and ω is some unknown compactly supported density function lying in the Orlicz–Sobolev space $L^P(\mathbb{C})$, where P will be chosen later. Note that

$$f_{\bar{z}} = \omega \in L^P(\mathbb{C}), \qquad f_z - 1 = S\omega \in L^P(\mathbb{C}), \tag{7.40}$$

where S is the Beurling–Ahlfors transform.

7.7.1. Results from Harmonic Analysis.

The proofs presented here of the theorems above exploit a number of substantial results in harmonic analysis. As a matter of fact, that is why we have chosen to present these important results in this way, using these new approaches which we hope will be useful in still more general settings and worthy of further exploration. The arguments also clearly illustrate the important rôle that the higher integrability properties of the Jacobians have to play. These were not recognized in earlier approaches. The critical exponent p_0 in Theorem 7.2 depends only on the constants in three inequalities which we now state.

The first is a direct consequence of [30].

THEOREM 7.8. (Coifman, Lions, Meyer, Semmes) *The Jacobian determinant $J(x, \phi)$ of a mapping $\phi \in W^{1,2}(\mathbb{C})$ belongs to the Hardy space $H^1(\mathbb{C})$ and we have the estimate*

$$\|J(x,\phi)\|_{H^1(\mathbb{C})} \leq C_1 \int_\mathbb{C} |D\phi|^2 \tag{7.41}$$

Next we have from [31]

THEOREM 7.9. (Coifman, Rochberg) *Let μ be a Borel measure in $\mathbb{C}$ such that its Hardy–Littlewood maximal function $M(x,\mu)$ is finite at a single point (and therefore at almost every point). Then $\log M(x,\mu) \in BMO(\mathbb{C})$ and its norm is bounded by an absolute constant,*

$$\|\log M(x,\mu)\|_{BMO} \leq C_2. \tag{7.42}$$

Finally we shall need the constant C_3 which appears in the H^1-BMO duality theorem of Fefferman, [35].

THEOREM 7.10. (Fefferman) *For $K \in BMO(\mathbb{C})$ and $J \in H^1(\mathbb{C})$ we have*

$$\left| \int K(x) J(x) \, dx \right| \leq C_3 \|K\|_{BMO} \|J\|_{H^1} \tag{7.43}$$

Note in advance that in our application of this theorem below the function J (Jacobian) will be continuous and so the exact meaning of the left hand side of (7.43) will not be a question (as it is in higher dimensions, see [54, 55]). There is one more constant that might be relevant to future developments, although it will not enter into our proofs. This is the constant Θ in the John-Nirenberg inequality. Having set these prerequisites in place we can now reveal that the exponent in Theorem 7.2 is

$$p_0 = 8C_1 C_2 C_3. \tag{7.44}$$

Before embarking on our proofs of the results discussed in §11 and §12 we point out that the approach presented here also covers the solution in the uniformly elliptic case, that is for $K(z,f) \leq K < \infty$. It shows that both f and its inverse belong to $W^{1,p}_{loc}(\mathbb{C},\mathbb{C})$ for some $p = p(K) > 2$ as first observed by Bojarski [21]. Our proof begins along the same lines as the classical proof so well presented in Ahlfors' book [3].

7.7.2. Existence for Exponential Distortion.
This subsection is devoted to the proof of Theorem 7.2.

As the distortion function $K(z) = \frac{1+|\mu(z)|}{1-|\mu(z)|}$ is equal to one outside the unit disk we may introduce the constant

$$A_p = \int_\mathbb{C} \left[e^{pK(z)} - e^p \right] dz < \infty \tag{7.45}$$

First we approximate μ by smooth functions via the mollification method. For each $0 < \nu \leq 1$ we define

$$\mu_\nu = \mu * \Phi_\nu \in C_0^\infty(2\mathbf{B}) \tag{7.46}$$

Clearly $|\mu_\nu| \leq |\mu| * \Phi_\nu < 1$, and by convexity of the function $t \mapsto \frac{1+t}{1-t}$ with $0 \leq t < 1$ we find that

$$\begin{aligned} K_\nu(z) &= \frac{1+|\mu_\nu|}{1-|\mu_\nu|} \leq \frac{1+|\mu|}{1-|\mu|} * \Phi_\nu \\ &= (K * \Phi_\nu)(z) \end{aligned} \tag{7.47}$$

Furthermore by convexity of the exponential function we have

$$\left[e^{pK_\nu(z)} - e^p \right] \leq \left[e^{pK(z)} - e^p \right] * \Phi_\nu$$

which provides us with the uniform bound

$$\int_{\mathbb{C}} \left[e^{pK_\nu(z)} - e^p \right] dz \le \int_{\mathbb{C}} \left[e^{pK(z)} - e^p \right] dz = A_p \tag{7.48}$$

Let us set
$$\gamma = \frac{p}{2}$$

This implies $\left[e^{\gamma K_\nu(z)} - e^\gamma \right] \in L^2(\mathbb{C})$. In fact

$$\int_{\mathbb{C}} \left[e^{\gamma K_\nu(z)} - e^\gamma \right]^2 \le \int_{\mathbb{C}} \left[e^{pK_\nu(z)} - e^p \right] \le A_p \tag{7.49}$$

Next, the maximal function of $e^{\gamma K_\nu(z)}$ is finite almost everywhere. Indeed

$$M(z, e^{\gamma K_\nu}) = e^\gamma + M(z, e^{\gamma K_\nu} - e^\gamma)) \tag{7.50}$$

and this last term is a constant plus a function in $L^2(\mathbb{C})$. Now consider the BMO functions

$$\mathcal{K}_\nu(z) = \frac{1}{\gamma} \log M(z, e^{\gamma K_\nu}) \tag{7.51}$$

By Theorem 7.9, the BMO norm of this function does not depend on ν,

$$\|\mathcal{K}_\nu\|_{BMO} \le \frac{2C_2}{p} \tag{7.52}$$

Moreover this function pointwise majorises the distortion function,

$$K_\nu(z) = \frac{1}{\gamma} \log \, e^{\gamma K_\nu(z)} \le \mathcal{K}_\nu(z)$$

In order to proceed we shall need a uniform L^2 bound for $\mathcal{K}_\nu$. For $t \ge e$ the function $t \mapsto \log^2(t)$ is concave, Accordingly

$$\begin{aligned}
\frac{1}{|2\mathbf{B}|} \int_{2\mathbf{B}} \mathcal{K}_\nu(z)^2 \, dz &= \frac{1}{4\gamma^2 |2\mathbf{B}|} \int_{2\mathbf{B}} \log^2 [M(z, e^{\gamma K_\nu})]^2 \\
&\le \frac{1}{4\gamma^2} \log^2 \frac{1}{|2\mathbf{B}|} \int_{2\mathbf{B}} [M(z, e^{\gamma K_\nu})]^2 \\
&\le \frac{1}{4\gamma^2} \log^2 \frac{1}{|2\mathbf{B}|} \int_{2\mathbf{B}} [M(z, e^{\gamma K_\nu} - e^\gamma) + e^\gamma]^2 \\
&\le \frac{1}{4\gamma^2} \log^2 \left[2e^{2\gamma} + \frac{2C}{|2\mathbf{B}|} \int_{\mathbb{C}} (e^{\gamma K_\nu} - e^\gamma)^2 \right]
\end{aligned}$$

Here we have used the L^2 inequality for the maximal operator and so C is an absolute constant. This combined with (7.49) yields

$$\int_{2\mathbf{B}} |\mathcal{K}_\nu(z)|^2 \, dz \le C_4 \log^2(1 + A_p) \tag{7.53}$$

where C_4 is an absolute constant. Let us now return to the mollified Beltrami equation,

$$f_{\bar{z}}^\nu = \mu_\nu(z) f_z^\nu \tag{7.54}$$

We look for a C^1-solution of (7.54) in the form

$$f_z^\nu(z) = e^{\sigma(z)}, \qquad f_{\bar{z}}^\nu(z) = \mu_\nu(z) e^{\sigma(z)} \tag{7.55}$$

where $\sigma \in W^{1,p_\nu}(\mathbb{C},\mathbb{C})$, for some $p_\nu > 2$, is compactly supported. Of course the necessary and sufficient condition for σ is that
$$(e^\sigma)_{\bar{z}} = (\mu_\nu e^\sigma)_z$$
or equivalently

(7.56) $$\sigma_{\bar{z}} = \mu_\nu \sigma_z + (\mu_\nu)_z$$

This equation is uniquely solved using the Beurling–Ahlfors transform, see §2. Note that $\sigma_z = S\sigma_{\bar{z}}$ and so equation (7.56) reduces to

(7.57) $$(\mathbf{I} - \mu_\nu S)\sigma_{\bar{z}} = (\mu_\nu)_z$$

As $\|S\|_2 = 1$ and as $\|\mu_\nu\|_\infty < 1$, there is $p_\nu > 2$ such that
$$\|\mu_\nu\|_\infty \|S\|_{p_\nu} < 1$$
In this case the operator $\mathbf{I} - \mu_\nu S$ has a continuous inverse. Thus

(7.58) $$\sigma_{\bar{z}} = (\mathbf{I} - \mu_\nu S)^{-1}(\mu_\nu)_z \in L^{p_\nu}(\mathbb{C})$$

and also

(7.59) $$\sigma_z = S\sigma_{\bar{z}} \in L^{p_\nu}(\mathbb{C})$$

Note that $\sigma_{\bar{z}}$ vanishes outside the support of μ_ν which is contained in $\mathbf{B}(0,2)$. Also $\sigma_z = S\sigma_{\bar{z}} = O(z^{-2})$ as $z \to \infty$. Thus $\sigma(z) \approx \frac{C}{z}$ asymptotically, for a suitable constant C. In fact of course

(7.60) $$\sigma(z) = (T\sigma_{\bar{z}})(z) = \frac{1}{2\pi i} \int_\mathbb{C} \frac{\sigma_{\bar{z}}(\zeta)}{\zeta - z} d\zeta \wedge d\bar{\zeta}$$

where T is the complex Riesz Potential. Hence σ is Hölder continuous with exponent $1 - \frac{2}{p_\nu}$, by the Sobolev Imbedding Theorem.

Now the solution f^ν of equation (7.55) is unique up to a constant as $f^\nu_{\bar{z}} = 0$ outside $2\mathbf{B}$ and as $f^\nu_z - 1 \in L^{p_\nu}(\mathbb{C})$. That is, f^ν is a principal solution to the Beltrami equation (7.54). It is important to realize here that the Jacobian of f^ν is strictly positive,

(7.61) $$J(z, f^\nu) = |f^\nu_z|^2 - |f^\nu_{\bar{z}}|^2 = (1 - |\mu_\nu|^2)|e^{2\sigma}| > 0$$

The Implicit Function Theorem tells us that f^ν is locally one-to-one. Another observation to make is that $\lim_{z\to\infty} f^\nu(z) = \infty$. It is an elementary topological exercise to show that $f^\nu : \mathbb{C} \to \mathbb{C}$ is a global homeomorphism of $\mathbb{C}$. Its inverse is C^1-smooth of course.

We now digress for a second to outline the existence proof in the classical setting where $K(z) \leq K < \infty$. As the sequence K_ν is uniformly bounded we find there is an exponent $p = p(K) > 2$ such that

(7.62) $$\|f^\nu_{\bar{z}}\|_p + \|f^\nu_z - 1\|_p \leq C_K$$

where C_K is a constant independent of ν. Hence the Sobolev Imbedding Theorem yields the uniform bound

(7.63) $$|f^\nu(a) - f^\nu(b)| \leq C_K |a-b|^{1-\frac{2}{p}} + |a-b|$$

The same inequality holds for the inverse map and hence

(7.64) $$|f^\nu(a) - f^\nu(b)| \geq \frac{|a-b|^{\frac{p}{p-2}}}{C_K + |a-b|^{\frac{2}{p-2}}}$$

We may assume that $f^\nu(0) = 0$. As the p-norms of $f_{\bar z}^\nu$ and $f_z^\nu - 1$ are uniformly bounded, we may assume that each converges weakly in $L^p(\mathbb{C})$ after possibly passing to a subsequence. From the uniform continuity estimates and Ascoli's Theorem, we may further assume $f^\nu \to f$ locally uniformly in $\mathbb{C}$. Obviously f satisfies the same modulus of continuity estimates and is therefore a homeomorphism. Moreover, it follows that the weak limits of $f_{\bar z}^\nu$ and $f_z^\nu - 1$ must in fact be equal to $f_{\bar z}$ and $f_z - 1$ respectively. Hence f is a homeomorphism in the Sobolev class $z + W^{1,p}(\mathbb{C})$, that is $f_{\bar z}$ and $f_z - 1$ in $L^p(\mathbb{C})$. Finally observe that $\mu_\nu \to \mu$ pointwise almost everywhere, and hence in $L^q(\mathbb{C})$, where q is the Hölder conjugate of p. The weak convergence of the derivatives shows that f is a solution to the Beltrami equation (7.37).

Now back to the more general setting. If we tried to follow the above argument we find the L^{p_ν} bounds are useless as we cannot keep them uniform. We therefore seek an alternative route via a Sobolev-Orlicz class where uniform bounds might be available. We begin with the elementary inequality

$$(7.65) \qquad (|u| + |v|)^2 \leq 2K(|u|^2 - |v|^2) + 4K^2|v - w|^2$$

whenever u, v, w are complex numbers such that $|w| \leq \frac{K-1}{K+1}|u|$ and $K \geq 1$. Indeed, $|v - w| \geq |v| - |w| \geq |v| - \frac{K-1}{K+1}|u|$, or equivalently

$$|u| + |v| \leq K(|u| - |v|) + (K+1)|v - w|$$

We multiply both sides of this inequality by $2(|u| + |v|)$ to find

$$\begin{aligned} 2(|u| + |v|)^2 &\leq 2K(|u|^2 - |v|^2) + 2(K+1)|v-w|(|u| + |v|) \\ &\leq 2K(|u|^2 - |v|^2) + (K+1)^2|v - w|^2 + (|u| + |v|)^2 \end{aligned}$$

from which inequality (7.65) is straightforward.

If we apply this inequality pointwise with

$$u = \phi_z^\nu, \quad v = \phi_{\bar z}^\nu, \quad w = \mu_\nu \phi_z^\nu$$

and $K = \mathcal{K}_\nu(z)$ as defined at (7.51), where

$$(7.66) \qquad \phi^\nu(z) = f^\nu(z) - z \in W^{1,2}(\mathbb{C}),$$

and use equations (7.54), (7.52) we can write

$$(|\phi_z^\nu| + |\phi_{\bar z}^\nu|)^2 \leq 2\mathcal{K}_\nu(|\phi_z^\nu|^2 - |\phi_{\bar z}^\nu|^2) + 4(\mathcal{K}_\nu)^2|\mu_\nu|^2$$

and hence

$$(7.67) \qquad |D\phi^\nu(z)|^2 \leq 2\mathcal{K}_\nu J(z, \phi^\nu) + 4|\mu_\nu \mathcal{K}_\nu|^2$$

Next we integrate this over the entire complex plane and use Theorems 7.8 and 7.10 to obtain

$$\begin{aligned} \int_\mathbb{C} |D\phi^\nu|^2 &\leq 2C_3 \|\mathcal{K}_\nu\|_{BMO} \|J(z, \phi^\nu)\|_{H^1} + 4 \int_{2\mathbf{B}} |\mathcal{K}_\nu|^2 \\ (7.68) \qquad &\leq \frac{4C_1 C_2 C_3}{p} \int_\mathbb{C} |D\phi^\nu|^2 + 4C_4 \log^2(1 + A_p) \end{aligned}$$

where in the latter step we have used the uniform bounds at (7.52) and (7.53).

It is clear at this point why we have chosen $p_0 = 8C_1 C_2 C_3$ at (7.44). The term $\int_\mathbb{C} |D\phi^\nu|^2$ in the right hand side can be absorbed in the left hand side. After doing this we obtain the uniform bounds in L^2

$$(7.69) \qquad \int_\mathbb{C} |D\phi^\nu|^2 \leq 8C_4 \log^2(1 + A_p)$$

7.7. EXISTENCE THEORY

which when unraveled reads as

$$\|D\phi^\nu\|_{L^2(\mathbb{C})} \leq C_5 \log\left(\int_\mathbf{B} e^{pK}\right) \tag{7.70}$$

This in turn leaves us with the local estimate for the mapping $f^\nu(z) = \phi^\nu(z) + z$, namely

$$\|Df^\nu\|_{L^2(\mathbf{B}_R)} \leq C_5 \left[R + \log\left(\int_\mathbf{B} e^{pK}\right)\right] \tag{7.71}$$

where $\mathbf{B}_R = \mathbf{B}(0, R)$. As f^ν is monotone (being a C^1 homeomorphism) we can apply the modulus of continuity estimate of Theorem 4.4,

$$|f^\nu(a) - f^\nu(b)| \leq C_6 \frac{R + \log\left(\int_\mathbf{B} e^{pK}\right)}{\log^{\frac{1}{2}}\left(e + \frac{R}{|a-b|}\right)} \tag{7.72}$$

for all $a, b \in \mathbf{B}_R$.

We need now to turn and consider the inverse map to f^ν. Let us denote it by $h^\nu = (f^\nu)^{-1} : \mathbb{C} \to \mathbb{C}$. As both f^ν and h^ν are smooth diffeomorphisms we find

$$\int_{\mathbf{B}_R} |Dh^\nu(w)|^2\, dw = \int_{h^\nu(\mathbf{B}_R)} K_\nu(z)\, dz$$
$$\leq \int_{h^\nu(\mathbf{B}_R)} K * \Phi_\nu(z)\, dz$$
$$\leq \int_{h^\nu(\mathbf{B}_R)} (K-1) * \Phi_\nu(z)\, dz + |h^\nu(\mathbf{B}_R)|$$
$$\leq \int_\mathbb{C} (K(z) - 1)\, dz + |h^\nu(\mathbf{B}_R)|$$
$$\leq C_p R^2 + \int_\mathbf{B} (K(z) - 1)\, dz$$

In this last inequality we have put in the uniform bound $|h^\nu(\mathbf{B}_R)| \leq C_p R^2$. One interesting way to see this estimate (though perhaps not the easiest) is via the Koebe distortion theorem for conformal mappings. We note that on $\mathbb{C} \setminus 2\mathbf{B}$ the functions f^ν are all univalent conformal mappings. At ∞ we have

$$f^\nu(z) = z + a_1 z^{-1} + a_2 z^{-2} + \cdots \tag{7.73}$$

and f^ν when restricted to $\mathbb{C} \setminus 2\mathbf{B}$ does not assume the value 0 (as $f^\nu(0) = 0$). Accordingly

$$|z| + \frac{4}{|z|} - 4 \leq |f^\nu(z)| \leq |z| + \frac{4}{|z|} + 4 \tag{7.74}$$

for all $|z| \geq 2$. In particular $\mathbf{B}(0, R-2) \subset f^\nu(\mathbf{B}(0, R)) \subset \mathbf{B}(0, R+6)$ whenever $R > 2$. This is equivalent to the inclusions

$$h^\nu(\mathbf{B}(0, R-2)) \subset \mathbf{B}(0, R) \subset h^\nu(\mathbf{B}(0, R+6))$$

which is more than enough to guarantee the estimate we used. Thus

$$\int_{\mathbf{B}(0,R)} |Dh^\nu|^2 \leq C\left(R^2 + \int_\mathbf{B} K\right) \tag{7.75}$$

and consequently we have the modulus of continuity estimate at Theorem 4.4 for h^ν using (4.33),

$$|h^\nu(x) - h^\nu(y)|^2 \leq \frac{CR^2 + \int_{\mathbf{B}} K}{\log\left(e + \frac{R}{|x-y|}\right)} \tag{7.76}$$

for $x, y \in \mathbf{B}(0, R)$. For f^ν this reads as

$$|f^\nu(a) - f^\nu(b)| \geq R \exp\left(\frac{-CR^2 - \int_{\mathbf{B}} K}{|a-b|^2}\right) \tag{7.77}$$

whenever $a, b \in \mathbf{B}(0, R)$ and $R \geq 1$. The uniform $W^{1,2}$ bounds, and the modulus of continuity estimates from above and below now enable us to pass to the limit. We find $f^\nu \to f$ and $h^\nu \to h = f^{-1}$ locally uniformly in $\mathbb{C}$ and Df^ν and Dh^ν converging weakly in $L^2_{loc}(\mathbb{C})$. As in the classical setting this implies that f is a homeomorphic solution to the Beltrami equation. Moreover $f_{\bar{z}}, f_z - 1 \in L^2(\mathbb{C})$ and the same is true for the inverse function

7.7.3. Uniqueness. We have already established the uniqueness of the principal solution in §8.2. However we note to the fundamental heuristic principle espoused by Schauder to the effect that the same estimates which are essential in establishing existence for the solutions of elliptic PDEs also provide uniqueness. We verify this principle in our setting next. Along the way we will be providing estimates which will be necessary in dealing with the more general existence and uniqueness theorem in the case of subexponential distortion.

Suppose we have two principal solutions f_1 and f_2 to the Beltrami equation at (7.37). Thus $\phi = f_1 - f_2$ satisfies the same equation and belongs to $W^{1,2}_{loc}(\mathbb{C})$. Inequality (7.67) reduces in this case to the pointwise estimate

$$|D\phi(z)|^2 \leq 2\mathcal{K}(z) J(z, \phi) \tag{7.78}$$

where $\mathcal{K}(z) = \frac{1}{\gamma} \log M(z, e^{\gamma K}) \in BMO(\mathbb{C})$. Upon integration, as before, we arrive at the inequality

$$\int_{\mathbb{C}} |D\phi|^2 \leq \frac{4C_1 C_2 C_3}{p} \int_{\mathbb{C}} |D\phi|^2 \tag{7.79}$$

which yields $D\phi = 0$ and hence $f_1 = f_2$ as desired. Thus normal solutions are unique. □

Now we want to study the factorization problem in a little greater depth.

COROLLARY 7.11. *Let $g \in W^{1,2}_{loc}(\Omega)$ be an arbitrary solution to equation (7.37), where $K(z) = \frac{1+|\mu(z)|}{1-|\mu(z)|}$ satisfies the exponential integrability condition (7.16), and let f be the principal solution of (7.37). Then there exists a holomorphic function $F : f(\Omega) \to \mathbb{C}$ such that*

$$g(z) = F(f(z)), \quad z \in \Omega \tag{7.80}$$

Proof. Obviously as f is a homeomorphism equation (7.80) uniquely determines F. The only remaining point is to show that F is holomorphic. Using notation introduced in the proof of Theorem 7.2 we factor g as

$$g(z) = F^\nu(f^\nu(z)) \tag{7.81}$$

where the F^ν are functions defined on $f^\nu(\Omega)$ by the rule $F^\nu = g \circ h^\nu$. As $h^\nu = (f^\nu)^{-1}$ is a diffeomorphism, each F^ν lies in $W^{1,2}_{loc}(f^\nu(\Omega))$. Fix an arbitrary disk U relatively compact in $f(\Omega)$. It suffices to show that $F|U$ is holomorphic. For all sufficiently small ν we see U is relatively compact in $f^\nu(\Omega)$ as $f^\nu \to f$ locally uniformly. We apply the chain rule to the smooth mappings f^ν to find

$$g_{\bar z} = F^\nu_w f^\nu_{\bar z} + F^\nu_{\bar w} \overline{f^\nu_z}$$
$$g_z = F^\nu_w f^\nu_z + F^\nu_{\bar w} \overline{f^\nu_{\bar z}}$$

Hence we have the following identities:

$$(g_{\bar z} - \mu_\nu g_z) f^\nu_z = F^\nu_{\bar w} J(z, f^\nu)$$
$$(g_z - \overline{\mu_\nu} g_{\bar z}) \overline{f^\nu_z} = F^\nu_w J(z, f^\nu)$$

We now integrate the absolute values of these equations and make the obvious change of variables in the right hand side:

$$\int_{h^\nu(U)} |g_{\bar z} - \mu_\nu g_z| |f^\nu_z| \, dz = \int_U |F^\nu_{\bar w}(w)| \, dw$$

$$\int_{h^\nu(U)} |g_z - \overline{\mu_\nu} g_z| |f^\nu_z| \, dz = \int_U |F^\nu_w(w)| \, dw$$

We then apply Hölder's inequality to obtain

(7.82) $$\int_U |F^\nu_{\bar w}(w)| \, dw \leq \|g_{\bar z} - \mu_\nu g_z\|_{L^2(\Omega)} \|f^\nu_z\|_{L^2(\Omega)} \leq C$$

(7.83) $$\int_U |F^\nu_w(w)| \, dw \leq \|g_z - \overline{\mu_\nu} g_{\bar z}\|_{L^2(\Omega)} \|f^\nu_z\|_{L^2(\Omega)} \leq C$$

where C is a constant independent of ν. Here we have replaced Ω by a slightly smaller relatively compact domain, which we continue to denote by Ω, in order to have the finite bounds at (7.82). It follows that the sequence $\{F^\nu\}$ is bounded in $W^{1,1}(U)$. The Compact Imbedding Theorem tells us that F^ν contains a subsequence converging strongly in $L^s(U)$ for all $1 \leq s < 2$. From the definition of F^ν we have $F = \lim_{\nu \to 0} F^\nu$ in $L^s(U)$. Clearly $F \in W^{1,1}(U)$. Also we note that $\{F^\nu_{\bar w}\}$ converges to 0 in $L^1(U)$. This is immediate from (7.82), the fact that $\mu_\nu \to \mu$ almost everywhere and $g_{\bar z} - \mu g_z = 0$. This shows that F is a solution in $W^{1,1}(U)$ to the Cauchy–Riemann equations,

(7.84) $$\frac{\partial F}{\partial \bar w} = 0 \quad a.e. \ U$$

Weyl's Lemma assures us that F is holomorphic in U which is what we wanted to prove. $\square$

There are two natural refinements of Corollary 7.11. The factorization of $W^{1,2}-$ solutions of the Beltrami equation holds under much weaker hypotheses on the distortion function K. Secondly, for K as in Theorem 7.2, but with possibly larger p, it is true that every solution in the Orlicz-Sobolev class $W^{1,Q}_{loc}$, with $Q(t) = t^2 \log^{-\alpha}(e+t)$, actually belongs to $W^{1,2}_{loc}$, and even to $W^{1,P}_{loc}$ with $P_\alpha(t) = t^2 \log^\alpha(e+t)$. This is what we now shall establish.

7.7.4. Critical Exponents.

To establish Theorem 7.3 concerning the jump in regularity it is clear we need only consider integer values of α. The case $\alpha = 0$ has just been handled above. We now proceed by induction on α.

In order that the principal solution given to us by Theorem 7.2 belongs to the Sobolev space $z + W^{1,P_\alpha}(\mathbb{C})$ with

$$P_\alpha(t) = t^2 \log^\alpha(e+t)$$

we must obtain uniform estimates in $L^2 \log^\alpha L$ for the functions ϕ^ν defined at (7.66). To do this we use (7.67) as a starting point. We recall

$$(7.85) \qquad |D\phi^\nu(z)|^2 \leq 2\mathcal{K}_\nu J(z, \phi^\nu) + 4|\mu_\nu \mathcal{K}_\nu|^2$$

Recall that $\phi^\nu \in C^1(\mathbb{C})$ and $|D\phi^\nu(z)| = O(|z|^{-2})$ at ∞.

The crucial fact used in the above proof was the inequality

$$(7.86) \qquad \begin{aligned} \int_\mathbb{C} \mathcal{K}_\nu J(z, \phi^\nu)\, dz &\leq C \|\mathcal{K}_\nu\|_{BMO} \|J(z, \phi^\nu)\|_{H^1} \\ &\leq \frac{C}{p} \int_\mathbb{C} |D\phi^\nu(z)|^2 dz \end{aligned}$$

We shall need a considerable generalization of this which we state as a lemma

LEMMA 7.12. *Let $1 \leq \mathcal{K} < \infty$ be a function in $BMO(\mathbb{C})$ with $\exp(\mathcal{K}) \in L^p(\mathbf{B})$ for some $p > 0$. Then for each $\alpha = 1, 2, 3 \ldots$ there is a constant C_α, depending only on α, such that*

$$(7.87) \qquad \begin{aligned} &\int_\mathbb{C} \mathcal{K}(z) J(z,\phi) \log^\alpha\left(e + \frac{|D\phi(z)|}{\|D\phi\|_2}\right) dz \\ &\leq C_\alpha \|\mathcal{K}\|_{BMO} \int_\mathbb{C} |D\phi(z)|^2 \log^\alpha\left(e + \frac{|D\phi(z)|}{\|D\phi\|_2}\right) dz \\ &\quad + C_\alpha \left(p + \int_\mathbf{B} e^{p\mathcal{K}}\right)^2 \int_\mathbb{C} |D\phi(z)|^2 \log^{\alpha-1}\left(e + \frac{|D\phi(z)|}{\|D\phi\|_2}\right) dz \end{aligned}$$

provided $\phi \in W^{1,P_\alpha}(\mathbb{C})$, $P_\alpha(t) = t^2 \log^\alpha(e+t)$.

The proof of this lemma is similar to that of (7.86) and presented with all its details in [62] and so we shall not prove it here.

The lemma obviously provides us with the tools to begin the induction. We begin by multiplying the distortion inequality at (7.85) by $\log^\alpha\left(e + \frac{|D\phi^\nu(z)|}{\|D\phi^\nu\|_2}\right)$ and

7.7. EXISTENCE THEORY

integrate this over the complex plane. We find

$$\int_{\mathbb{C}} |D\phi^\nu|^2 \log^\alpha \left(e + \frac{|D\phi^\nu|}{\|D\phi^\nu\|_2}\right)$$

$$\leq 2\int_{\mathbb{C}} \mathcal{K}_\nu J(z,\phi^\nu) \log^\alpha \left(e + \frac{|D\phi^\nu|}{\|D\phi^\nu\|_2}\right)$$

$$+ 4\int_{\mathbf{B}} \mathcal{K}_\nu^2 \log^\alpha \left(e + \frac{|D\phi^\nu|}{\|D\phi^\nu\|_2}\right)$$

$$\leq 2C_\alpha \|\mathcal{K}_\nu\|_{BMO} \int_{\mathbb{C}} |D\phi^\nu|^2 \log^\alpha \left(e + \frac{|D\phi^\nu|}{\|D\phi^\nu\|_2}\right)$$

$$+ 2C_\alpha \left(p + \int_{\mathbf{B}} e^{p\mathcal{K}_\nu}\right)^2 \int_{\mathbb{C}} |D\phi^\nu|^2 \log^{\alpha-1} \left(e + \frac{|D\phi^\nu|}{\|D\phi^\nu\|_2}\right)$$

$$+ 4\int_{\mathbf{B}} \mathcal{K}_\nu^2 \log^\alpha \left(e + \frac{|D\phi^\nu|}{\|D\phi^\nu\|_2}\right)$$

As $\|\mathcal{K}_\nu\|_{BMO} \leq 2C_2 p^{-1}$, the first term can be absorbed in the left hand side if we choose p sufficiently large. Such a choice of p depends only on α. After making such a choice we arrive at the inequality,

$$\int_{\mathbb{C}} |D\phi^\nu|^2 \log^\alpha \left(e + \frac{|D\phi^\nu|}{\|D\phi^\nu\|_2}\right)$$

$$\leq 4C_\alpha \left(p + \int_{\mathbf{B}} e^{p\mathcal{K}_\nu}\right)^2 \int_{\mathbb{C}} |D\phi^\nu|^2 \log^{\alpha-1} \left(e + \frac{|D\phi^\nu|}{\|D\phi^\nu\|_2}\right)$$

$$+ 4\int_{\mathbf{B}} \mathcal{K}_\nu^2 \log^\alpha \left(e + \frac{|D\phi^\nu|}{\|D\phi^\nu\|_2}\right)$$

We have already established the necessary uniform bounds for the L^2–norms of the sequence $\{D\phi^\nu\}$, thus the last term remains bounded as $\nu \to \infty$.

The second term remains bounded by the induction hypothesis (where we notice that the factor $\int_{\mathbf{B}} e^{p\mathcal{K}_\nu}$ remains bounded as we assume that e^K is locally integrable with a sufficiently large power).

Now passing to the limit as $\nu \to \infty$ we conclude that $\phi \in W^{1,P_\alpha}(\mathbb{C})$ which puts our solution f in the desired class.

Having established the $L^2 \log^\alpha L$ estimates for Df^ν independent of ν we now can apply a duality argument to see the jump in regularity we discussed. Suppose that

$$g \in W^{1,Q_\alpha}_{loc}(\Omega), \qquad Q_\alpha = t^2 \log^{-\alpha}(e+t)$$

Following the argument at (7.81) we factor g as

$$g(z) = F^\nu(f^\nu(z))$$

for functions $F^\nu : f^\nu(\Omega) \to \mathbb{C}$. As we let $\nu \to \infty$ we have $g = F(f(z))$, $F : f(\Omega) \to \mathbb{C}$, and we want to show that F is in fact holomorphic. We obtain the following version of (7.82) using Hölder's inequality in Orlicz spaces

$$(7.88) \qquad \int_U |F^\nu_{\bar{w}}(w)|\, dw \leq C_\alpha \|g_{\bar{z}} - \mu_\nu g_z\|_{Q_\alpha} \|f^\nu_z\|_{P_\alpha} \leq C$$

$$(7.89) \qquad \int_U |F^\nu_w(w)|\, dw \leq C_\alpha \|g_z - \overline{\mu_\nu} g_{\bar{z}}\|_{Q_\alpha} \|f^\nu_{\bar{z}}\|_{P_\alpha} \leq C$$

where C is a constant independent of ν. It then follows that F is holomorphic in $f(\Omega)$ and consequently the solution enjoys the additional regularity we claimed.

7.7.5. Existence for Subexponential Distortion. In this subsection we prove Theorem 7.6 The idea of the proof is to look for a solution in the form

$$f(z) = f^2(f^1(z)) \tag{7.90}$$

where f^1 and f^2 are principal solutions of certain related Beltrami equations which satisfy the hypotheses of Theorem 7.2. At first glance it seems rather surprising that one can make both the distortion functions of f^1 and f^2 exponentially integrable, with large exponents. However the composition of $W^{1,2}$-solutions need not be a $W^{1,2}$-mapping. A degree of integrability of Df is lost. Let us first discuss how to achieve this factorization at the level of the Beltrami coefficient μ whose distortion $K(z) = \frac{1+|\mu(z)|}{1-|\mu(z)|} < \infty$ almost everywhere. Suppose we factor K arbitrarily as

$$K(z) = K_2(z) \, K_1(z) \tag{7.91}$$

where $1 \leq K_1(z), K_2(z) \leq K(z)$ almost everywhere. To each such factorization there corresponds the associated two Beltrami equations. First, if possible, solve

$$f^1_{\bar{z}}(z) = \mu_1(z) f^1_z(z) \tag{7.92}$$

where $\mu_1(z) = \frac{K_1(z)-1}{K_1(z)+1} \frac{\mu(z)}{|\mu(z)|}$. Here, when $\mu(z) = 0$ we understand the indefinite quotient to be equal to 1. At such points z we have $\mu_1(z) = 0$ as $1 \leq K_1(z) \leq K(z) = 1$. Having solved the equation (7.92) for a homeomorphism $w = f^1(z)$ we then define the Beltrami coefficient for $f^2 = f^2(w)$ by the rule

$$\mu_2(w) = \frac{\mu(z) - \mu_1(z)}{1 - \overline{\mu_1(z)}\mu(z)} \frac{f^1_z}{\overline{f^1_z}} \tag{7.93}$$

and then we look for a normal solution, if such possibly exists, to the equation

$$f^2_{\bar{w}}(w) = \mu_2(w) f^2_w(w) \tag{7.94}$$

The formula for the distortion function of f^2, denoted by $\mathcal{K}_2 = \mathcal{K}_2(w)$, reads as

$$\mathcal{K}_2(w) = \frac{1+|\mu_2(w)|}{1-|\mu_2(w)|} = K_2(z) \tag{7.95}$$

where $w = f^1(z)$. A purely formal application of the chain rule reveals that $f(z) = f^2 \circ f^1$ solves the original Beltrami equation $f_{\bar{z}}(z) = \mu(z) f_z(z)$.

We have several things to check in order that we can carry out this procedure. The first is to find an appropriate factorization of $K(z)$. We do this as follows;

$$K_1(z) = \frac{K(z)}{\log(e - 1 + K^{\frac{1}{p}}(z)) - \log\log(e - 1 + K^{\frac{1}{p}}(z))} \tag{7.96}$$

where p is the exponent of Theorem 7.2 It is not difficult to see that the denominator ranges between 1 and $K^{\frac{1}{p}} \leq K$ as we desired. Using the elementary inequality $x - \log x \geq \frac{e-1}{e} x$ for $x \geq 1$ we find that

$$K_1(z) \leq \frac{2pK(z)}{\log(e + K(z))} \tag{7.97}$$

Thus

$$\int_{\mathbf{B}} \exp(pK_1(z)) \, dz \leq \int_{\mathbf{B}} \exp\left(\frac{2p^2 K}{\log(e+K)}\right) < \infty \tag{7.98}$$

as we now set $p_* = 2p^2$. Of course $K_1 = 1$ outside the unit disk as $K = 1$ there. Now Theorem 7.2 tells us that there is in fact a unique principal solution $w = f^1(z)$ in $z + W^{1,2}(\mathbb{C})$ to the equation at (7.92). Next we consider the Beltrami equation (7.94) for $f^2 = f^2(w)$. The distortion function takes the form at $w = f^1(z)$,

$$\mathcal{K}_2(w) = \mathcal{K}_2(z) = \log(e - 1 + K^{\frac{1}{p}}) - \log\log(e - 1 + K^{\frac{1}{p}})$$

Hence

(7.99) $$p\mathcal{K}_2(w) \leq 2p + \log K(z) - \log\log(e - 1 + K(z))$$

Next, integration by substitution yields

$$\int_{f^1(\mathbf{B})} e^{p\mathcal{K}_2(w)} \, dw = \int_{\mathbf{B}} \exp[p\mathcal{K}_2(f^1(z))] \, J(z, f^1) dz$$

$$\leq e^{2p} \int_{\mathbf{B}} \frac{K(z)}{\log(e - 1 + K(z))} J(z, f^1) dz$$

(7.100) $$\leq e^{2p} \int_{\mathbf{B}} \frac{K}{\log(e + K)} J(z, f^1) dz$$

We also notice $\mathcal{K}_2(w) = 1$ outside of the set $f^1(\mathbf{B})$.

Up to this point, we have simply used routine inequalities. Now we must use something substantially deeper. Recall Theorem 7.2 has actually told us that $f^1 \in z + W^{1,2}(\mathbb{C})$. Therefore its Jacobian is $L \log L$ integrable according to Theorem 5.6. Although it is not essential to the proof, we record the precise bounds

$$\int_{\mathbf{B}} J(z, f^1) \log\left(e + \frac{J(z, f^1)}{J_{\mathbf{B}}}\right) \leq C \int_{2\mathbf{B}} |Df^1(z)|^2 \, dz$$

$$\leq C \log \int_{\mathbf{B}} e^{pK_1(z)} \, dz$$

(7.101) $$\leq C \log \int_{\mathbf{B}} \exp\left(\frac{2p^2 K}{\log(e + K)}\right)$$

There is a point to make regarding the change of variables $w = f^1(z)$. We recall that from the construction of f^1 in Theorem 7.2 it is a limit of C^1–diffeomorphisms which are uniformly bounded in $W^{1,2}_{loc}(\mathbb{C}, \mathbb{C})$ and whose distortion functions converge to $K_1(z)$. The usual limiting argument shows that the integration by substitution is legitimate.

Returning to the estimate, we use the elementary inequality $xJ \leq se^x + J\log(e + J/s)$ to obtain

$$\int_{f^1(\mathbf{B})} e^{p\mathcal{K}_2(w)} dw$$

$$\leq e^{2p} J_{\mathbf{B}} \int_{\mathbf{B}} \exp\left(\frac{K}{\log(e + K)}\right) + C e^{2p} \int_{\mathbf{B}} J(z, f^1) \log\left(e + \frac{J(z, f^1)}{J_{\mathbf{B}}}\right) dz$$

(7.102) $$C \int_{\mathbf{B}} \exp\left(\frac{K}{\log(e + K)}\right) \log\left[\int_{\mathbf{B}} \exp\left(\frac{2p^2 K}{\log(e + K)}\right)\right]$$

Thus $\mathcal{K}_2$ possesses the necessary degree of integrability to apply Theorem 7.2. Accordingly, there exists a unique principal solution $f^2 \in w + W^{1,2}(\mathbb{C})$ to the Beltrami equation (7.94). That the composition $f = f^2 \circ f^1$ again satisfies the original Beltrami equation is yet another simple matter of an approximation argument based on the weak convergence of the derivatives. We should point out here that f does not lie in the Sobolev class $W_{loc}^{1,2}$. It is in $W_{loc}^{1,1}(\mathbb{C})$ and its Jacobian is locally integrable. To find the precise Orlicz–Sobolev class containing f we use the inequality

$$\frac{|Df|^2}{\log(e+|Df|^2)\log\log(3+|Df|^2)} \leq \frac{KJ}{\log(e+KJ)\log\log(3+KJ)}$$
$$\leq C_1 J + C_2 \exp\left(\frac{2p^2 K}{\log(e+K)}\right)$$

where C_1 and C_2 are constants depending only on p. Hence $|Df|$ belongs to the Orlicz–Sobolev space $W_{loc}^{1,P}(\mathbb{C})$ where

(7.103) $$P(t) = \frac{t^2}{\log(e+t)\log\log(3+t)}$$

The example we gave at (7.22) shows we cannot expect to improve this regularity result any further. The modulus of continuity estimates at equation (7.29) are straightforward consequences of (7.17) when applied to f^1 and f^2.

Next, it might seem that we could set up an induction to further improve Theorem 7.6 by factoring a Beltrami coefficient in such a way that both the factors have distortion satisfying the hypotheses of Theorem 7.6. That is just how we got Theorem 7.6 from Theorem 7.2. However, the very delicate point concerns the change of variables. Theorem 7.2 provided us with a $W_{loc}^{1,2}$ map f^1 whose Jacobian was in $L\log L$. Theorem 7.6 does not give us so good a map. Although the Jacobian will be locally integrable (we have a homeomorphism), there seems no natural dual class where we might put the distortion function $\mathcal{K}_2$ so as to even be able to make sense of the integral that will appear as that at (7.102). Despite this problem, we do not need p_* in Theorem 7.6 to be large. The existence assertion remains valid for arbitrary positive ϵ in place of p_*. To see this we consider the factorization

$$K(z) = \frac{p_*}{\epsilon} \cdot \frac{\epsilon K(z)}{p_*}$$

Then $K_1(z) = \frac{\epsilon K(z)}{p_*}$ satisfies the hypotheses of Theorem 7.6,

$$\int_{\mathbf{B}} \exp\left(\frac{p_* K_1}{\log(e+K_1)}\right) \leq C \int_{\mathbf{B}} \exp\left(\frac{\epsilon K}{\log(e+K)}\right) < \infty$$

and the factor $K_2(z) = \frac{p_*}{\epsilon} \geq 1$ represents a quasiconformal mapping. In this way we establish the existence of a solution by composing the mapping provided for us by Theorem 7.6 with distortion $K_1(z)$ and the quasiconformal map. Notice that the quasiconformal map will lie in the space $W_{loc}^{1,s}(\mathbb{C})$ for all

$$s < \frac{2p_*}{p_* - \epsilon}$$

by Astala's Theorem, yet the composition will not lie in $W_{loc}^{1,2}(\mathbb{C})$ in general.

7.8. Global Solutions

In this section we give a fairly general existence and uniqueness theorem for the Beltrami equation without any assuming that μ is compactly supported. We point out that without any condition on the distortion function at ∞ we cannot hope to guarantee that a homeomorphic solution in the entire plane $\mathbb{C}$ extends continuously to the Riemann sphere $\overline{\mathbb{C}}$.

7.8.1. Solutions on $\mathbb{C}$.

THEOREM 7.13. *Let $\mu : \mathbb{C} \to \mathbb{B}$ be a measurable function valued in the unit disk and suppose that the distortion function of the associated Beltrami equation $h_{\bar{z}} = \mu h_z$ has the property that*

$$\exp\left(\frac{K}{1 + \log K}\right) \in L^p_{loc}(\mathbb{C}) \tag{7.104}$$

for some $p > 0$. Then there is a continuous solution $f : \mathbb{C} \to \mathbb{C}$ to the Beltrami equation with the following properties:

- *f is injective, but not necessarily onto.*
- *$f \in W^{1,R}_{loc}(\mathbb{C})$ with*

$$R(t) = \frac{t^2}{\log(e+t) \log\log(3+t)}$$

- *If $g \in W^{1,R}_{loc}(\mathbb{C})$ is any other solution, then there is a holomorphic function $\phi : f(\mathbb{C}) \to \mathbb{C}$ such that $g(z) = \phi \circ f(z)$.*

It is to be noted that locally, say on a bounded open set U, the solution we prove the existence of enjoys the same regularity properties as the principal solution of the equation whose Beltrami coefficient is $\mu \chi_U$.

Proof. For every $n = 1, 2, \ldots$ we set

$$\mu_n(z) = \begin{cases} \mu(z) & |z| \leq n \\ 0 & |z| > n \end{cases}$$

As μ_n has a distortion function $K_n(z) \leq K(z)$ with $\exp\left(\frac{K_n}{1+\log K_n}\right) \in L^p(\mathbf{B}_n)$ there is a unique principal solution $h_n : \mathbb{C} \to \mathbb{C}$ to the Beltrami equation

$$\bar{\partial} h_n = \mu_n(z) \partial h_n, \qquad h_n(z) = z + o(1)$$

We normalise this solution by setting

$$f_n(z) = \frac{h_n(z) - h_n(0)}{h_n(1) - h_n(0)} \tag{7.105}$$

We want to extract from this sequence a subsequence converging locally uniformly to a mapping $f : \mathbb{C} \to \mathbb{C}$ with $f(0) = 0$ and $f(1) = 1$. It suffices to show that for each disk $\mathbf{B} = \mathbf{B}(0, R)$, $R \geq 2$, there is a subsequence converging locally uniformly on $\mathbf{B}$, for then the usual diagonal argument will apply to generate the sequence we seek. Let us fix such a disk $\mathbf{B}$. Next let $h : \mathbb{C} \to \mathbb{C}$ be the principal solution to the equation

$$\bar{\partial} h = \mu(z) \chi_\mathbf{B}(z) \partial h$$

which we further normalise so that $h(0) = 0$ and $h(1) = 1$. The factorization theorem tells us that for each n, with $n \geq R$, the map f_n can be written as

$$f_n(z) = \phi_n(h(z)) \tag{7.106}$$

where $\phi_n : h(\mathbf{B}) \to \mathbb{C}$ form a family of conformal mappings with $\phi_n(0) = 0$ and $\phi_n(1) = 1$. Therefore, each ϕ_n when restricted to the domain $h(\mathbf{B}) \setminus \{0, 1\}$ omits the three values $0, 1, \infty$ and so the family $\{\phi_n\}$ is normal by Montel's Theorem. Thus every sequence admits a subsequence which converges locally uniformly on $h(\mathbf{B})$ and this limit map is either a conformal mapping or a constant by Hurwitz's Theorem. Examining the uniform convergence on small circles about 0 and 1 and using the fact that each ϕ_n is continuous and bounded on $h(\mathbf{B})$ quickly shows there are no constant limits. Moreover, the removable singularity theorem tells us that any conformal limit has a conformal extension to $h(\mathbf{B})$. Therefore $\{\phi_n\}$ is normal on $h(\mathbf{B})$ with every limit being a conformal map.

This process provides us with our limit map on $\mathbf{B}$, namely

$$\lim_{k\to\infty} f_{n_k} = \lim_{k\to\infty} \phi_{n_k}(h(z)) = \phi(h(z)) = f(z)$$

where $\phi_{n_k} \to \phi$, a conformal mapping, locally uniformly on $h(\mathbf{B})$.

Next we need to observe the uniform bounds on the integrals of the Jacobians,

(7.107) $$\int_U J(z, f_{n_k})\,dz \leq |f_{n_k}(U)| \leq C_U$$

for every relatively compact subset U of $\mathbf{B}$. The constant C_U does not depend on n_k as f_{n_k} is uniformly bounded on compact subsets of $\mathbf{B}$. With the aid of the elementary inequality

$$\frac{|Df_{n_k}|^2}{\log(e+|Df_{n_k}|^2)\log\log(3+|Df_{n_k}|^2)}$$
$$\leq \frac{K(z)J(z,f_{n_k})}{\log(e+K(z)J(z,f_{n_k}))\log\log(3+K(z)J(z,f_{n_k}))}$$
$$\leq C_p J(z,f_{n_k}) + C_p \exp\left(\frac{pK(z)}{1+\log K(z)}\right)$$

we conclude that the sequence $\{Df_{n_k}\}$ is bounded in $L^R(U)$ for every relatively compact subdomain U of $\mathbf{B}$. Thus $f_{n_k} \to f$ weakly in $W^{1,R}(U)$. In particular, $f \in W^{1,R}_{loc}(\mathbf{B})$. Moreover

(7.108) $$f_{\bar{z}} - \mu f_z = \lim_{k\to\infty}(\bar{\partial}f_{n_k} - \mu(z)\partial f_{n_k}) = 0,$$

the limit being the weak limit in $W^{1,R}_{loc}(\mathbf{B})$. In particular, f solves the Beltrami equation. This completes the proof of the theorem. $\square$

Do note here that although each f_{n_k} is a homeomorphism of $\mathbb{C}$ onto $\mathbb{C}$, the limit map f need not be. As an example, consider the map f constructed in Theorem 3.1 for given weight function $\mathcal{A}$. We easily extend f conformally to $\mathbb{C}$ by linearly extending ρ. Then f is continuous at ∞ with the definition $f(\infty) = \infty$. Now the mapping of finite distortion

$$g(z) = \frac{1}{f(\frac{1}{z})}, \qquad g(0) = 0$$

is defined and continuous on $\mathbb{C}$ but it is bounded. Near ∞ we have $K(z,g) = K(\frac{1}{z}, f)$. Whence

(7.109) $$\int_{\mathbb{C}} \exp(\mathcal{A}(K(z,g)))\frac{dz}{1+|z|^4} = C + \int_{\mathbf{B}} \exp(\mathcal{A}(K(z,f)))\,dz < \infty$$

Recall that $\mathcal{A}(K) = \frac{K}{1+\log^{1+\epsilon} K}$ for $\epsilon > 0$ is a sufficient weight function for the boundedness of g. However it is not if $\epsilon = 0$. This gives us a clue as to a sharp condition for the existence of a global solution on $\overline{\mathbb{C}}$.

In fact it is probable that this example proscribes the precise limits for the existence of bounded entire mappings of finite distortion (or, if one prefers, solutions to the Beltrami equation).

7.8.2. Solutions on $\overline{\mathbb{C}}$. What we want to do now is to give a condition that will imply the solution defined above can be extended continuously, and therefore homeomorphically to $\overline{\mathbb{C}}$, by setting $f(\infty) = \infty$. One could achieve this by demanding uniform asymptotic estimates on the integrability properties of K to give modulus of continuity bounds as is done in [**26, 97**]. However, this approach does not provide the regularity estimates we want and is in fact stronger than is necessary. We will simply assume subexponential integrability of the distortion in a spherical sense.

THEOREM 7.14. *Let $\mu : \overline{\mathbb{C}} \to \mathbf{B}$ be a measurable function and suppose that the distortion function of the associated Beltrami equation $h_{\bar{z}} = \mu h_z$ has the property that*

$$(7.110) \qquad \int_{\mathbb{C}} \exp\left(\frac{pK}{1+\log K}\right) \frac{dz}{1+|z|^4} < \infty$$

for some $p > 0$. Then there is a homeomorphic solution $f : \overline{\mathbb{C}} \to \overline{\mathbb{C}}$ to the Beltrami equation such that

$$(7.111) \qquad \int_{\mathbb{C}} \frac{|\Psi(z)|^2}{\log(e+\Psi(z))\log\log(3+\Psi(z))} \frac{dz}{1+|z|^4} < \infty$$

where $\Psi(z) = \frac{1+|z|^2}{1+|f(z)|^2}|Df(z)|$.

Notice the example at the end of §14.1 shows the condition (7.110) cannot be weakened to

$$(7.112) \qquad \int_{\mathbb{C}} \exp\left(\frac{pK}{1+\log^{1+\epsilon} K}\right) \frac{dz}{1+|z|^4} < \infty$$

for any positive ϵ in order to guarantee a continuous extension to $\mathbb{C}$.

Of course it follows from our factorisation results that this solution is unique up to the normalization $f(0) = 0$, $f(1) = 1$, and $f(\infty) = \infty$. Further note that if

$$\limsup_{z \to \infty} \frac{|z|}{|f(z)|} < \infty,$$

then (7.111) reduces to the weighted L^R estimate for $|Df|$ with weight given by the spherical measure $\frac{d\zeta}{1+|\zeta|^4}$. In general, the weight depends on the growth rate of f near ∞. Bounds on the rate can be provided by local Hölder or modulus of continuity estimates at ∞. However, when formulated in these terms the result appears not to be sharp.

Proof. Let f be the normalized injective solution given by Theorem 7.13. We first want to show that f has a well defined limit at ∞. To this end set

$$(7.113) \qquad g(z) = \frac{1}{f(\frac{1}{z})} : \mathbb{C} \setminus \{0\}$$

We compute that

$$(7.114) \qquad \mu_g(z) = \mu(\frac{1}{z})\left(\frac{z}{|z|}\right)^4$$

The distortion functions are therefore related by the formula $K_g(z) = K_f(\frac{1}{z})$. We then have

$$\int_{\mathbf{B}} \exp\left(\frac{pK_g}{1+\log K_g}\right) dz = \int_{\mathbb{C}\setminus\mathbf{B}} \exp\left(\frac{pK_f}{1+\log K_f}\right) \frac{d\zeta}{|\zeta|^4}$$

$$\leq 2\int_{\mathbb{C}} \exp\left(\frac{pK_f}{1+\log K_f}\right) \frac{d\zeta}{1+|\zeta|^4} < \infty$$

Therefore we can find a unique principal solution to the equation

$$h_{\bar{z}} = \mu_g(z)\chi_{\mathbf{B}}(z)h_z$$

defined on $\mathbb{C}$. Now on the domain $\mathbf{B} \setminus \{0\}$ both g and h are embeddings satisfying the same Beltrami equation. We apply the factorization theorem to see

$$g(z) = (\varphi \circ h)(z), \qquad z \in \mathbf{B} \setminus \{0\}$$

Now

(7.115) $$\varphi : h(\mathbf{B} \setminus \{0\}) \to g(\mathbf{B} \setminus \{0\})$$

is conformal. Therefore it admits a continuous extension over the point $h(0)$, possibly this value is ∞. This gives a well defined limit to g at 0, and therefore a well defined limit to f at ∞. As f is a homeomorphism on $\mathbb{C}$ it must be the case that $f(\infty) = \infty$.

This implies now that $h \in W^{1,R}_{loc}(\mathbf{B})$,

$$R(t) = \frac{t^2}{\log(e+t)\log\log(3+t)}$$

Next, as $\frac{1}{f(z)} = h(\frac{1}{z})$ we have

$$|Dh(z)| = |Df(1/z)|\frac{1}{|z|^2|f(\frac{1}{z})|^2}$$

and the result claimed follows from the change of variables formula.

7.9. Holomorphic Dependence

Here we present an interesting rigidity phenomenon. We show that if μ_λ is a holomorphically varying family of Beltrami coefficients (defined below) with initial distortion in the exponential class, then all distortions lie in the exponential class and there is a holomorphically varying family of solutions. However the singular points, where $|\mu(z)| = 1$ remain fixed and the holomorphic variation trivializes to factor through a quasiconformal holomorphic variation. First we take a little time to recount a bit of the theory.

7.9.1. Holomorphic Motions. Recently there has been a third route discovered to the theory of quasiconformal mappings in the plane which has had many important consequences. The notion appeared in a paper of Mane–Sad–Sullivan, [78] and has been dubbed holomorphic motions. Basically the idea is that a holomorphic perturbation of the identity in the space of injections $A \hookrightarrow \mathbb{C}$ of a set $A \subset \mathbb{C}$ is necessarily a quasiconformal mapping. Here is a precise definition.

Let $\mathbf{B}$ denote the open unit disk in $\mathbb{C}$. A holomorphic motion of a set $A \subset \overline{\mathbb{C}}$ is a map $f : \mathbf{B} \times A \to \overline{\mathbb{C}}$ such that

(1) for each fixed $z \in A$, the function $\lambda \to f(\lambda, z)$ is holomorphic in $\mathbf{B}$,

(2) for each fixed $\lambda \in \mathbf{B}$, the map $f_\lambda : A \to \overline{\mathbb{C}}$, $f_\lambda(z) = f(\lambda, z)$, is an injection,
(3) the mapping f_0 is the identity on A.

Note that there is no assumption regarding the continuity of f as a function of z or the pair (λ, z). That such continuity occurs is a consequence of the following remarkable λ–lemma of Mañé–Sad–Sullivan [78]. We give here the result as extended by Slodkowski [100]; see also [11, 34]. Lempert has communicated a particularly straightforward proof of this result, using only the solution to the $\overline{\partial}$ equation, to the second author.

THEOREM 7.15. *If $f : \mathbf{B} \times A \to \overline{\mathbb{C}}$ is a holomorphic motion of $A \subset \mathbb{C}$, then f has an extension to $F : \mathbf{B} \times \overline{\mathbb{C}} \to \overline{\mathbb{C}}$ such that*
 (i) F is a holomorphic motion of $\overline{\mathbb{C}}$,
 (ii) F is continuous in $\mathbf{B} \times \overline{\mathbb{C}} \to \overline{\mathbb{C}}$,
 (iii) $F_\lambda : \overline{\mathbb{C}} \to \overline{\mathbb{C}}$ is K-quasiconformal with $K \leq \frac{1+|\lambda|}{1-|\lambda|}$ for each $\lambda \in \mathbf{B}$.

Holomorphic motions arise naturally in the study of complex dynamical systems, for example the iteration of polynomials or rational functions in the complex plane. This is because as one holomorphically varies the parameters of the dynamical system, (for example the coefficients of the polynomials), periodic cycles and their eigenvalues also vary analytically. Density results then imply that the stable and unstable regions vary analytically. Using the dynamics one can place restrictions on the parameters for the system so that this analytic variation is actually a holomorphic motion. One then can apply Theorem 7.15 to conclude that the variation of the fixed points induces a holomorphic motion of the whole complex plane which commutes with the dynamics. This is how quasiconformal mappings arise naturally when one discusses conjugacy (topological equivalence after a change of coordinates) between "nearby" dynamical systems.

7.9.2. Dependence on Parameters. We now discuss the holomorphic dependence on parameters in a much weaker setting than classically. We hope that this provides some impetus to using the more general theory of mappings of finite distortion in dynamical applications.

The reader should have no trouble verifying the following corollary. At each stage of our existence proof all the maps are seen to depend holomorphically on the parameters.

COROLLARY 7.16. *Let $\mu(z, \zeta)$ be a function defined on $\mathbb{C} \times \mathbf{B}$, measurable with respect to z and analytic with respect to ζ. Suppose that $\mu(z, \zeta) = 0$ if $|z| > 1$, and*

$$K_\zeta(z) = \frac{1 + |\mu(z, \zeta)|}{1 - |\mu(z, \zeta)|} \in L^Q(\mathbf{B})$$

Then there exist normal solutions $f(z, \zeta) \in W^{1,P}_{loc}(\mathbb{C})$ to the Beltrami equation

$$f_{\overline{z}}(z, \zeta) = \mu(z, \zeta) f_z(z, \zeta)$$

depending holomorphically on the parameter ζ. That is $f(z, \cdot) : \mathbf{B} \to \mathbb{C}$ is holomorphic for each $z \in \mathbb{C}$.

Here P and Q are Orlicz functions for which we have global existence as discussed in §14. For instance if $K_\zeta \in Exp(\mathbf{B})$ we have $f \in W^{1,P}_{loc}(\mathbb{C})$ with $P = t^2/\log(e+t)$ and so forth.

Typically one might consider a holomorphic variation of a Beltrami coefficient μ with $\|\mu\|_\infty = 1$ such as
$$\mu(z, \lambda) = \lambda \mu(z)$$
However, such variations are largely uninteresting, as in this case $\|\mu(z, \lambda)\|_\infty = |\lambda| < 1$ and we have classical solutions. Moreover, our original μ is attained only at the boundary. However, an example of a non-trivial variation might be one of the form
$$\mu(z, \lambda) = \left(\frac{\lambda - \mu}{1 - \bar{\mu}\lambda}\right)^m$$
Such variations have the property that if $|\mu(z_j)| \to 1$, then $|\mu(z_j, \lambda)| \to 1$ for all λ. This is of course a general phenomena. Indeed the maximum principle shows that if $|\mu(z_0, \lambda_0)| = 1$ for some λ_0 in the unit disk, then $\mu(z_0, \lambda) \equiv \mu(z_0, \lambda_0)$. Thus the singularity set is preserved under a holomorphic variation of the Beltrami equation. Moreover, if
$$\mu(z, \zeta) : \mathbb{C} \times \mathbf{B} \to \mathbf{B}$$
is a holomorphic variation of $\mu = \mu(\cdot, 0)$, then the Schwarz lemma implies that $\mu(z, \cdot)$ is a contraction in the hyperbolic metric $\rho_\mathbf{B}(z, w)$. That is

(7.116) $$\rho_\mathbf{B}(\mu(z, \lambda), \mu(z, 0)) \leq \rho_\mathbf{B}(0, \lambda) = \log \frac{1 + |\lambda|}{1 - |\lambda|}$$

In particular, this easily implies that $|\mu(z_j)| \to 1$ if and only if $|\mu(z_j, \lambda)| \to 1$.
Since
$$\rho_\mathbf{B}(\mu(z, \lambda), \mu(z, 0)) \geq \rho_\mathbf{B}(\mu(z, \lambda), 0) - \rho_\mathbf{B}(0, \mu(z, 0))$$
we have

(7.117) $$\frac{1 + |\mu(z, \lambda)|}{1 - |\mu(z, \lambda)|} \frac{1 + |\mu(z, 0)|}{1 - |\mu(z, 0)|} \leq \frac{1 + |\lambda|}{1 - |\lambda|}$$

if this difference is positive. While the result is trivial otherwise. This now reads as

(7.118) $$K(z, \lambda) \leq \frac{1 + |\lambda|}{1 - |\lambda|} K(z, 0)$$

where $K(z, \lambda)$ is the distortion function of $\mu(z, \lambda)$. The estimate at (7.118) implies that if $K(z) = K(z, 0)$ is exponentially integrable, then so too is $K(z, \lambda)$. Similarly for other classes, such as those for which
$$\int_\mathbf{B} e^{pK \log^{-1} K} < \infty$$
for some $p > 0$.

7.9.3. Rigidity. Finally, here is a fairly strong rigidity theorem which the reader should have no trouble establishing for themselves given the sketch below. It shows that holomorphic variations of unbounded Beltrami coefficients are in a sense trivial. They reduce to holomorphic variations of associated bounded coefficients.

THEOREM 7.17. *Let $\mu(z, \lambda)$ be a holomorphically varying family of Beltrami coefficients such that $\mu_0(z) = \mu(0, \lambda)$ has distortion function K_0 and*

(7.119) $$\exp(K_0) \in L^p(\mathbf{B})$$

for some $p > 0$. Let $\{f_\lambda : \mathbb{C} \to \mathbb{C}\}$ be the holomorphically varying family of principal solutions. Then there is a holomorphically varying family of quasiconformal mappings $\{g_\lambda : \mathbb{C} \to \mathbb{C}\}_{\lambda \in \mathbf{B}}$ with $g_0 = identity$ such that

$$f_\lambda = g_\lambda \circ f_0$$

Sketch of Proof. The idea is to factor the distortion functions into a bounded part times the initial unbounded distortion function exactly as in the argument following (7.90) in §11. Inequality (7.118) tells us that there is a holomorphically varying family of Beltrami coefficients whose distortion functions $M_\lambda(z)$ are such that

- $M_0(z) = 1$
- $M_\lambda(z)K(z,0) = K(z,\lambda)$
- $M_\lambda(z) \leq \frac{1+|\lambda|}{1-|\lambda|}$

We now of course choose $\{g_\lambda\}$ to be the holomorphically varying family of quasiconformal principal solutions. The only obstacle really is that we cannot apply the chain rule to deduce that $g_\lambda \circ f_0$ are also principal solutions to the same Beltrami equation as f_λ. For then uniqueness would apply and give us $g_\lambda \circ f_0 = f_\lambda$ as desired. Actually all we need for this observation is that $g_\lambda \circ f_0 \in z + W^{1,Q}(\mathbb{C})$, $Q(t) = t^2 \log^{-1}(e+t)$. However even that is not obvious. What we must do is go back to the proof following (7.90) and carry out the entire procedure again, although it is a little easier with bounded distortion. It is clear we retain the uniform estimates from this factorization for each λ. It follows that $g_\lambda \circ f_0$ lies in the correct Sobolev class and we are done. □

It is fairly obvious that this result holds in more generality. For instance the Beltrami coefficients need only be defined in a planar domain Ω, and also they hold if the variation is actually in a subexponential class. We leave it to the reader to develop these generalities.

7.10. Examples and Non-Uniqueness

In this section we give a number of examples to show the regularity theorems we have proven for solutions to the Beltrami equations are optimal. In fact the result is easily fashioned in all dimensions, however, for the sake of clarity here, we leave this generalization to the reader. Here we will be particularly concerned with the radial stretching, defined for any positive number k by

(7.120) $$f(z) = \frac{z}{|z|} \log^{-k} \frac{e}{|z|}$$

The map f is defined in the unit disk $\mathbf{B}$, maps this disk to itself and is the identity on the boundary. We find that near the origin,

(7.121) $$|Df(x)|^2 = |z|^{-2} \log^{-2k} \frac{e}{|z|}$$

and

(7.122) $$J(z,f) = k|z|^{-2} \log^{-2k-1} \frac{e}{|z|}$$

The distortion function is therefore

(7.123) $$K(z,f) = \frac{1}{k} \log \frac{e}{|z|}.$$

Indeed, this formula is valid as soon as
$$|z| \leq e^{1-k}$$
We compute that
$$f_z = \frac{1}{2}\left(\frac{z}{\bar{z}}\right)^{\frac{1}{2}} \log^{-k} \frac{e^2}{|z|^2}\left(\frac{1}{2z} + \frac{k}{z}\log^{-1}\frac{e^2}{|z|^2}\right)$$
and that
$$f_{\bar{z}} = \frac{1}{2}\left(\frac{z}{\bar{z}}\right)^{\frac{1}{2}} \log^{-k} \frac{e^2}{|z|^2}\left(\frac{-1}{2\bar{z}} + \frac{k}{z}\log^{-1}\frac{e^2}{|z|^2}\right)$$
to find that
$$(7.124) \qquad \mu_f(z) = \frac{z}{\bar{z}} \frac{k - \log\frac{e}{|z|}}{k + \log\frac{e}{|z|}}$$

We now point out some important features of this map. First, the Jacobian, $J(z,f)$, is always integrable. Secondly, the map
$$(7.125) \qquad f \in W^{1,2}(\mathbf{B}, \mathbb{C}) \quad \text{if and only if } 2k > 1$$
Thirdly, the distortion function $K(z,f)$ is exponentially integrable,
$$(7.126) \qquad \int_{\mathbf{B}} e^{pK(z,f)} dz\, d\bar{z} < \infty$$
if and only if $p < 2k$.

7.10.1. Quasireflections. Before proving the main result of this section (a modification and improvement of an example first given by us in [57]) we need to make a few preliminary remarks and formulate and prove a couple of auxiliary results.

Let us first fix $k \geq 0$. The k–quasireflection in the circle $\mathbb{S}(a,r)$, $r < 1$, is defined as
$$(7.127) \qquad \Psi_{a,r}(z) = a + r\frac{z-a}{|z-a|}\log^k\frac{er}{|z-a|}$$
From the identities of §3 we find the complex dilatation $\mu_{a,r}$ of $\Psi_{a,r}$ to be
$$(7.128) \qquad \mu_{a,r}(z) = -\frac{z-a}{|z-a|}\frac{\log\frac{er}{|z-a|} - k}{\log\frac{er}{|z-a|} + k}, \qquad z \in \mathbf{B}(a,r)$$
In particular
$$(7.129) \qquad |\mu_{a,r}(a)| = 1, \quad \text{and} \quad |\mu_{a,r}(z)| = \frac{|k-1|}{|k+1|}, \quad |z-a| = r$$
The distortion function is then
$$(7.130) \qquad K_{a,r}(z) = \max\left\{\frac{1}{k}\log\frac{er}{|z-a|}, k\log^{-1}\frac{er}{|z-a|}\right\}$$
We compute the differential of $\Psi_{a,r}$ and find it to be
$$D\Psi_{a,r}(z) = \frac{r}{|z-a|}\log^k\frac{er}{|z-a|}\left[\mathbf{I} + \left(1 + k\log^{-1}\frac{er}{|z-a|}\right)\frac{(z-a) \otimes (z-a)}{|z-a|^2}\right]$$
On the disk $\mathbf{B}(a,r)$ we have the inequality
$$(7.131) \qquad |\Psi_{a,r}(z) - z| \leq r\log^k\frac{er}{|z-a|}$$

as $|\Psi_{a,r}(z) - z| \leq |\Psi_{a,r}(z) - a|$. which follows directly from the definition. This inequality implies that for each $q > 0$ we have

(7.132) $$\int_{\mathbf{B}(a,r)} |\Psi_{a,r}(z) - z|^q \, dz \leq C(k,q) \, r^q \, |\mathbf{B}(a,r)|$$

where $C(k,q) = 2^{q+1} \max\{1, e^{1-kq}(kq)^{kq}\}$. Indeed

$$\int_{\mathbf{B}(a,r)} |\Psi_{a,r}(z) - z|^q \, dz \leq r^q \int_{\mathbf{B}(a,r)} \log^{kq} \frac{e\,r}{|z-a|} \, dz$$
$$\leq 2\pi r^q \int_0^r t \log^{kq} \frac{e\,r}{t} \, dt$$
$$\leq C(k,q) r^q |\mathbf{B}(a,r)|$$

where we have used the fact that for $0 \leq t \leq r$,

$$t \log^{kq} \frac{e\,r}{t} \leq \max\{r, re^{1-kq}(kq)^{kq}\} = rC(k,q)$$

To see this,

$$\frac{d}{dt} t \log^{kq} \frac{e\,r}{t} = \log^{kq} \frac{e\,r}{t} - kq \log^{kq-1} \frac{e\,r}{t}$$
$$= \log^{kq} \frac{e\,r}{t}(1 - kq \log^{-1} \frac{e\,r}{t})$$

Thus $t \log^{kq} \frac{e\,r}{t}$ has its maximum on the interval $[0, r]$ either at the endpoint $t = r$ or at $t = re^{1-kq}$, and this maximum value is either r or $re^{1-kq}(kq)^{kq}$.

Next we seek to find the Orlicz–Sobolev space in which $\Psi_{a,r}$ lies.

We compute the norm of $D\Psi_{a,r}(x)$ from the eigenvalues of $D^t\Psi_{a,r}(z)D\Psi_{a,r}(z)$.

$$\text{Trace}[D^t\Psi_{a,r}(z)D\Psi_{a,r}(z)] = \frac{r^2}{t^2} \log^{2k} \frac{e\,r}{t}\left(1 + k^2 \log^{-2} \frac{e\,r}{t}\right)$$

with $t = z - a$. From this we deduce

(7.133) $$\frac{1}{2}\frac{r}{t} \log^k \frac{e\,r}{t} \leq |D\Psi_{a,r}(z)| \leq (1+k)\frac{r}{t} \log^k \frac{e\,r}{t}$$

Hence $D\Psi_{a,r}(z) \notin L^2_{loc}(\mathbf{B}(a,r))$. However, we have the estimate

(7.134) $$\frac{|Df|^2}{\log^\alpha(e+|Df|^2)} \leq C(k,\alpha) \frac{r^2}{|z-a|^2} \log^{2k-\alpha} \frac{e\,r}{|z-a|}$$

Hence

$$\int_{\mathbf{B}(a,r)} \frac{|Df|^2}{\log^\alpha(e+|Df|^2)} \leq C(k,\alpha) \int_{\mathbf{B}(a,r)} \frac{r^2}{|z|^2} \log^{2k-\alpha} \frac{e\,r}{|z|} dz$$
$$= C(k,\alpha) \, 2\pi r^2 \int_0^r \log^{2k-\alpha} \frac{e\,r}{t} \frac{dt}{t}$$
$$= C(k,\alpha) \, |\mathbf{B}(a,r)| \int_1^\infty s^{2k-\alpha} ds$$

(7.135) $$= C(k,\alpha) \, |\mathbf{B}(a,r)| \frac{1}{2k-\alpha+1}$$

provided

(7.136) $$2k + 1 < \alpha$$

Thus $D\Psi_{a,r} \in L^P(\mathbf{B}(a,r))$ for

(7.137) $$P(t) = t^2 \log^{-\alpha}(e+t), \qquad \alpha > 2k+1$$

Next, as $\Psi_{a,r}(z) - z$ vanishes on $\partial\mathbf{B}(a,r)$, we obtain, upon using integration by parts,

(7.138) $$\int_{\mathbf{B}(a,r)} D^t\eta(z)[\Psi_{a,r}(z) - z]\, dz = -\int_{\mathbf{B}(a,r)} [D^t\Psi_{a,r} - \mathbf{I}]\eta(z)\, dz$$

for any test mapping $\eta \in C^\infty(\mathbb{C})$ (there is no problem near 0 here). This shows that $\Psi_{a,r} - \mathbf{I}$ when restricted to $\mathbf{B}$ belongs to the Sobolev space $W^{1,P}(\mathbf{B})$ for $P(t)$ satisfying (7.137).

We now construct a map by piecing together these quasireflections in a careful way. Given a domain $\Omega \in \mathbb{C}$, an *exact packing* of Ω by disks is an infinite family $\mathcal{F} = \{\mathbf{B}_j\}_{j=1}^\infty$ of disjoint open disks $\mathbf{B}_j \subset \Omega$ such that

$$\left|\Omega \setminus \bigcup_{j=1}^\infty \mathbf{B}_j\right| = 0$$

The existence of such an exact packing for any domain Ω follows from Vitali's covering lemma. To simplify the estimates involved, we shall require that the disks all have radius bounded by 1.

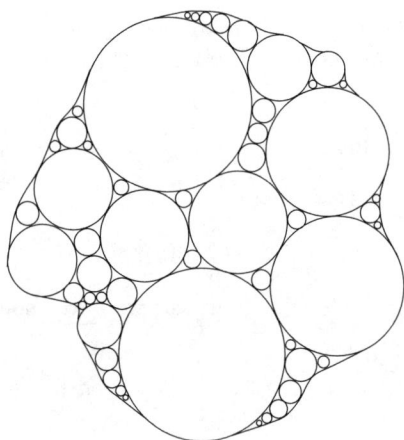

An exact packing

Let us assume, simply for convenience, that $\Omega \subset \mathbb{C}$ is a domain and $|\Omega| < \infty$. Let $\mathcal{F} = \{\mathbf{B}_j\}_{j=1}^\infty$, $\mathbf{B}_j = \mathbf{B}(a_j, r_j)$, be an exact packing of Ω by disks of radius $r_j \leq 1$. We fix $k > 0$ and let

$$\Psi_j = \Psi_{a_j, r_j}$$

be the quasireflection in the circles $\partial \mathbf{B}_j$. We define a function F_k in the following piecewise fashion:

(7.139) $$F_k(z) = \begin{cases} \Psi_j(z) & \text{if } z \in \mathbf{B}_j \\ z & \text{otherwise} \end{cases}$$

7.10. EXAMPLES AND NON-UNIQUENESS

The inequality at (7.132) immediately implies that

$$\int_\Omega |F_k(z) - z|^q \, dz = \sum_{j=1}^\infty \int_{\mathbf{B}_j} |\Psi_j(z) - z|^q \, dz$$
$$\leq C(k,q) \sum_{j=1}^\infty |\mathbf{B}_j| = C(k,q) |\Omega| < \infty$$

for every $1 < q < \infty$. Thus $F_k \in L^q(\Omega)$. We next define the matrix function A by

(7.140) $$A(z) = \begin{cases} D\Psi_j(z) & \text{if } z \in \mathbf{B}_j \\ \mathbf{I} & \text{otherwise} \end{cases}$$

From the inequality at (7.135) we have, as above,

$$\int_\Omega \frac{|A(z)|^2}{\log^\alpha(e + |A(z)|)} \, dz \leq C(n,k)|\Omega| < \infty$$

for every $\alpha > 2k + 1$. Therefore to prove that $F_k \in W^{1,P}(\Omega)$, with P as at (7.137) we need only verify that $A(z)$ coincides with $DF_k(z)$ in the sense of distributions. To do this we choose a test mapping $\eta \in C_0^\infty(\Omega)$ and apply (7.138) to Ψ_j using integration by parts to see

$$\int_\Omega D^t \eta(z) F_k(z) \, dz = \int_\Omega D^t \eta(z)[F_k(z) - z] \, dz + \int_\Omega D^t \eta(z) z \, dz$$
$$= \sum_{j=1}^\infty \int_{\mathbf{B}_j} D^t \eta(z)[\Psi_j(z) - z] \, dz - \int_\Omega \eta(z) \, dz$$
$$= -\sum_{j=1}^\infty \int_{\mathbf{B}_j} [D^t \Psi_j(z) - z] \eta(z) \, dz - \int_\Omega \eta(z) \, dz$$
$$= -\sum_{j=1}^\infty \int_{\mathbf{B}_j} D^t \Psi_j(z) \eta(z) \, dz$$
$$= -\int_\Omega A^t(z) \eta(z) \, dz$$

It now follows that $F_k \in W^{1,P}(\Omega, \mathbb{C})$. It is easy to see that $J(z, F_k) \leq -1$ almost everywhere in Ω. To get an orientation preserving mapping, we simply set

(7.141) $$h_k(z) = \overline{F_k(z)}$$

We now consider the integrability properties of the distortion function. From (7.130) we see that on the disk $\mathbf{B}(a_j, r_j)$

(7.142) $$K(z, h_k) = K(z, \Psi_{a_j, r_j}) = \max\left\{ \frac{1}{k} \log \frac{e\, r_j}{|z - a_j|}, \; k \log^{-1} \frac{e\, r_j}{|z - a_j|} \right\}$$

The reader may now observe that we need only consider the integrability properties of the function

(7.143) $$K(z) = \begin{cases} \frac{1}{k} \log \frac{e\, r_j}{|z - a_j|} & \text{if } z \in \mathbf{B}_j \\ 1 & \text{otherwise} \end{cases}$$

While this is not the distortion function when it is less than 1 that region is immaterial to our considerations. The pointwise distortion inequality

(7.144) $$|Dh_k(z)|^2 \leq K(z) J(z, h_k)$$

holds in each ball $\mathbf{B}_j$ by construction, and therefore almost everywhere in Ω. Let $\alpha > 0$, then

$$
\begin{aligned}
\int_\Omega e^{\alpha K(z)} dz &= \sum_{j=1}^\infty \int_{\mathbf{B}_j} e^{\alpha K(z)} dz \\
&= 2\pi \sum_{j=1}^\infty \int_0^{r_j} e^{\alpha \frac{1}{k} \log \frac{e\, r_j}{|z-a|}} dz \\
&= 2\pi \sum_{j=1}^\infty \int_0^{r_j} \left(\frac{e r_j}{t}\right)^{\frac{\alpha}{k}} t\, dt \\
&= 2\pi e^{\alpha/k} r_j^{\alpha/k} \sum_{j=1}^\infty \int_0^{r_j} t^{1-\frac{\alpha}{k}} dt \\
&= \frac{2\pi k}{2k-\alpha} e^{\alpha/k} \sum_{j=1}^\infty r_j^2 \\
&= \frac{2k}{2k-\alpha} e^{\alpha/k} |\Omega| < \infty
\end{aligned}
$$

where, in order to perform the integration, we must have

(7.145) $$\alpha < 2k$$

In summary, the function h_k has the following properties:

(1) $h_k \in W^{1,P}(\Omega, \mathbb{C})$ for $P(t) = t^2 \log^{-\alpha}(e+t)$, for all $\alpha > 2k+1$, thus
(2) $h_k \in W^{1,p}(\Omega, \mathbb{C})$ for all $p < 2$.
(3) $\exp[K(z, h_k)] \in L^\alpha(\Omega)$ if and only if $\alpha < 2k$.

Let us define another function,

(7.146) $$g_k(z) = \begin{cases} a_j + r_j f\left(\frac{z-a_j}{r_j}\right) & \text{if } z \in \mathbf{B}_j \\ z & \text{otherwise} \end{cases}$$

where f is the radial stretching defined at (7.120). The map g_k has the property that it maps Ω homeomorphically onto itself.

Following the arguments above, and noting (7.125), we see that g_k has the following properties:

(1) $g_k \in W^{1,2}(\Omega, \mathbb{C})$ if $2k > 1$,
(2) $\exp[K(z, h_k)] \in L^\alpha(\Omega)$ if and only if $\alpha < 2k$.

Indeed, g_k and h_k differ on the disk $\mathbf{B}_j$ by composition with an inversion. In particular

(7.147) $$\mu_{g_k}(z) = \mu_{h_k}(z), \qquad \text{a.e. } \Omega$$

and hence $K(z, g_k) = K(z, h_k)$. We set $\mu_k(z) = \mu_{g_k}(z)$. More explicitly, for $z \in \mathbf{B}_j$ we have

(7.148) $$\mu_k(z) = -\frac{z-a_j}{|z-a_j|} \frac{\log \frac{e\, r_j}{|z-a_j|} - k}{\log \frac{e\, r_j}{|z-a_j|} + k},$$

We now obtain the following theorem which shows that although we have existence of solutions to the Beltrami equation, uniqueness is not to be expected outside of the space $W^{1,2}(\Omega)$, nor are properties such as openness and discreteness.

THEOREM 7.18. Let $\Omega \subset \mathbb{C}$ be a domain. For each $k > \frac{1}{2}$, the Beltrami coefficient μ_k has the following properties. First, the distortion function
$$K_k(z) = \frac{1 + |\mu_k(z)|}{1 - |\mu_k(z)|}$$
is locally exponentially integrable,

(7.149) $$\int_E e^{pK_k(z)} dz < \infty$$

for all E relatively compact in Ω and all $p < 2k$. Second, the Beltrami equation
$$f_{\bar{z}} = \mu_k(z) f_z$$
admits two solutions. The first is a homeomorphism $g : \Omega \to \Omega$ with
$$g \in W^{1,P}_{loc}(\Omega), \qquad P(t) = t^2 \log^\beta(e+t), \quad \beta < 2k - 1$$
and the second is $h : \Omega \to \overline{\mathbb{C}}$ which is not a homeomorphism, but
$$h \in W^{1,Q}_{loc}(\Omega), \qquad Q(t) = t^2 \log^{-\alpha}(e+t), \quad \alpha > 2k + 1$$
In particular, $g \in W^{1,2}_{loc}(\Omega)$ and $h \in W^{1,s}_{loc}(\Omega)$ for all $s < 2$. These solutions distinguish themselves by the fact that there is no meromorphic function $\phi : \Omega \to \mathbb{C}$ such that

(7.150) $$h_k = \phi \circ g_k$$

Note too, that the mapping constructed is certainly not open or discrete. Yet it has finite distortion and lies in a relatively nice Sobolev space.

7.11. Equations in the Plane

A complex function $f(z) = u(x,y) + iv(x,y)$ in the Sobolev class $W^{1,2}_{loc}(\Omega, \mathbb{R}^2)$ is K–quasiregular if

(7.151) $$u_x^2 + u_y^2 + v_y^2 + v_x^2 \leq \left(K + \frac{1}{K}\right)(u_x v_y - u_y v_x)$$

This defines quasiregularity in terms of the Hilbert-Schmidt norm, and (7.151) can be rewritten as

(7.152) $$\text{Trace}(D^t f \, Df) \leq \left(K + \frac{1}{K}\right) \det Df$$

We have mentioned earlier that the most general quasilinear system of first order PDEs for a mapping $f : \Omega \to \Omega'$ takes the form

(7.153) $$f_{\bar{z}} = \mu_1(z,f) f_z + \mu_2(z,f) \overline{f_z}$$

where the ellipticity condition reads as

(7.154) $$|\mu_1(z,f)| + |\mu_2(z,f)| \leq k < 1$$

In matrix notation we write

(7.155) $$D^t f(z) H(z,f) Df(z) = J(z,f) G(z,f)$$

Here $G : \Omega \times \mathbb{C} \to \mathbb{R}^{2\times 2}$ and $H : \Omega' \times \mathbb{C} \to \mathbb{R}^{2\times 2}$ are measurable functions valued in the space of positive definite matrices of determinant 1. An explicit relation between distortion tensors (as G and H are usually called) and the complex coefficients μ_1 and μ_2 are

(7.156) $$\mu_1 = \frac{G_{11} - G_{22} + 2i\, G_{12}}{G_{11} + G_{22} + H_{11} + H_{22}}, \quad \mu_2 = \frac{H_{22} - H_{11} - 2i\, H_{12}}{G_{11} + G_{22} + H_{11} + H_{22}}$$

As we have seen in this paper, various regularity properties of mappings with finite distortion depend on the degree of integrability of the distortion function. One could choose to impose conditions on G or on H, depending on a particular application. Each choice would yield different properties of solutions.

7.11.1. Fluid Flow. The connections between second order elliptic equations in the plane and elliptic systems of first order PDEs is well known and well understood. It is appropriate to recall here the pioneering work on gas dynamics and planar fluid flow of a number of authors, and presented in Bers' [**18**]. Also the related theory of compressible fluid flow [**23**].

Without getting into technicalities, both theories are concerned with the *continuity equation*

(7.157) $$[A(|\nabla \theta|)\theta_x]_x + [A(|\nabla \theta|)\theta_y]_y = 0$$

for the potential function $\theta = \theta(x, y)$ of a two dimensional flow. In the case of subsonic flows the complex gradient $f(z) = \theta_x - i\theta_y$ is a quasiregular mapping. However the uniform ellipticity is lost when one approaches a critical value of velocity. The theory of mappings of unbounded (or finite) distortion seems to be useful in these critical cases and offer a way to studying these equations. Let us formulate this following [**18**].

The (planar) flow of a perfect fluid in a region Ω is described by giving the density ρ and the velocity components u, v as functions of the cartesian coordinates x, y and time t. Usually the steady state equations are studied so there is no dependence on t. A complete description also requires two other thermodynamic variable, such as pressure p and temperature T. Typically one deals with the adiabatic and isentropic flows in which pressure is a definite function of density. For an ideal gas $p = c\rho^\gamma$, where $\gamma > 1$ is the ratio of specific heats, but more general pressure density relations are important. The derivative $\frac{dp}{d\rho}$ is positive and denoted c^2. It is interpreted as the speed of propagation of small disturbances in a flow and is referred to as the *local speed of sound*. By Bernoulli's theorem for potential flows, density is a (decreasing) function of the speed q of the flow. We have

$$c^2 = -\frac{q\rho}{\dot{\rho}(q)}, \quad M^2 = -\frac{q}{\rho}\frac{d\rho}{dq}$$

The type of quasilinear second order equation one studies depends on the Mach number M. The equation is elliptic at subsonic speeds ($M < 1$) and hyperbolic at supersonic speeds. Typically one considers "nice" subsonic flows with perhaps regions where the flow becomes supersonic at a "shock" boundary. Thus avoiding the delicate issues of degeneracy of the equations. The potential equation takes the form

(7.158) $$(c^2 - u^2)\phi_{xx} - 2uv\phi_{xy} + (c^2 - v^2)\phi_{yy}, \qquad u = \theta_x, v = \theta_y$$

The continuity equation reads

$$(\rho u)_x + (\rho v)_y = 0$$

so there is a "stream function" such that $\rho u = \psi_y$ and $\rho v = \psi_x$. The velocity potential ϕ and the stream function ψ are coupled by the equations

(7.159) $$\rho \phi_x = \psi_y, \quad \rho \phi_y = -\psi_x$$

Notice that for constant density (say $\rho = 1$) corresponding to the case of an incompressible flow, we have the Cauchy–Riemann equations. These equations quickly imply

$$\phi_x \psi_y - \phi_y \psi_x = \frac{1}{\rho}(\psi_x^2 + \psi_y^2)$$
$$\phi_x \psi_y - \phi_y \psi_x = \rho(\phi_x^2 + \phi_y^2)$$

from which we deduce that the mapping $F = (\phi, \psi)$ has

$$\phi_x^2 + \phi_y^2 + \psi_x^2 + \psi_y^2 = (\rho + \frac{1}{\rho})(\phi_x \psi_y - \phi_y \psi_x)$$

giving us the distortion relation

$$\|DF\|^2 = (\rho + \frac{1}{\rho})J(z, F)$$

and so F is quasiregular. Thus our results allow one to study this equation in the critical cases of infinite (or zero) pressure. The fundamental result here is that the factorization theorem tells us that an L^2 solution is of the form $f(g(z))$ where g is a homeomorphic solution and f is analytic. Thus, after a change of coordinates, the flow is the same as that for an incompressible flow (an analytic function). Thus we know the topological nature of stream lines and so forth of these non-standard various flows about objects.

One of the most remarkable and useful techniques developed for the study of non-linear equations is the hodographic transformation. While the idea seems to go back at least to Gauss, its utility in this area is amply demonstrated in the work of Lavrentiev and Bers. Roughly speaking, given a system of first order PDEs and a solutions $w = f(z)$ to this system, the hodographic transformation asks us to simply regard w as an independent variable in the hodograph plane, and z as a function of w. If the original system is quasilinear, this simple trick converts the equation into a linear system, with variable coefficients.

If we study our potential equation in the hodograhic plane we see similar features. We consider an irrotational flow, so we have

$$u_y = v_x$$

as well as the equation (7.158). This quickly lead us to

$$c^2(u_x^2 + u_y^2) = (c^2 - v^2)(uv_x + vv_y)^2$$
$$c^2(v_x^2 + v_y^2) = (c^2 - u^2)(vu_x + uu_y)^2$$

where $F = (u, v)$. Hence the Hilbert–Schmidt norm of the differential satisfies

(7.160) $(c^2 - u^2 - v^2)\|DF\|^2 = (2c^2 - u^2 - v^2)J(z, F) - (uv_y - vu_x)^2 - (uu_x - vu_y)^2$

so that

(7.161) $$\|DF\|^2 \leq \left(1 + \frac{c^2}{c^2 - u^2 - v^2}\right) J(z, F)$$

and once again F is quasiregular, with the same consequences as above. Here the distortion function is

$$K = \left(1 + \frac{c^2}{c^2 - u^2 - v^2}\right)$$

This time the critical case occurs when we approach the speed of sound. Indeed more appears here, we must also have

$$(uv_y - vu_x)^2 + (vu_x - vu_y)^2 \to 0$$

on the critical set $(c^2 = u^2 + v^2)$. Thus $\nabla h = 0$ where $h = \frac{u}{v}$, so we can identify the critical set as a level line of h.

7.11.2. Variational Interpretations. The existence theorem for the Beltrami equation has an interesting variational interpretation in much the same way as the Riemann mapping problem has for the Cauchy–Riemann equations. To see this we consider two simply connected planar domains Ω and Ω'. We wish to study the problem of deforming Ω into Ω' using the least possible energy. Such problems arise frequently in elasticity theory. In our model, the energy of a deformation $f : \Omega \to \mathbb{C}$ is measured by the functional

(7.162) $$\mathcal{E}[f] = \text{Trace} \int_\Omega Df(x) G^{-1}(x) D^t f(x)\, dx$$

where $G : \Omega \to \mathbb{R}^{2 \times 2}$ is a given measurable metric tensor on Ω (which we think of as describing the mechanical and physical properties of the material from which Ω is made) such that we have the ellipticity bounds

$$\frac{1}{K}|\xi|^2 \leq \langle G(z)\xi, \xi \rangle \leq K|\xi|^2, \qquad \text{a.e. } \Omega, \ \xi \in \mathbb{R}^2, \ 1 \leq K < \infty$$

Here $\det G(z) = 1$. In elasticity theory the integrand

$$E(z, A) = \text{Trace}(AG^{-1}(z)A^t)$$

is called the *stored energy* function.

We now try to minimise the energy functional at (7.162) over all Sobolev mappings in $W^{1,2}(\Omega, \mathbb{R}^2)$ which essentially cover Ω', more precisely we assume that the image of very subset of full measure in Ω contains a subset of full measure in Ω'. The class of Sobolev mappings is used for compactness reasons. There is no reason to suppose, and it will often not be the case, that the minima are smooth. The energy of such an f can easily be estimated from below using a change of variables,

$$\mathcal{E}[f] = \int_\Omega \text{Trace}[Df(x)G^{-1}(x)D^t f(x)]\, dx$$

(7.163) $$\geq 2 \int_\Omega |J(x,f)|\, dx$$

(7.164) $$\geq 2|\Omega'|$$

Where we have used Hadamard's inequality:

$$\text{Trace}(AG^{-1}A) \geq 2\,|\det A|$$

as $\det G = 1$. Notice that equality occurs here if and only if $A^t A = |\det A|G$. Therefore if we want equality at (7.163) the mapping $f : \Omega \to \Omega'$ must solve the Beltrami system

(7.165) $$D^t f(x) Df(x) = J(x,f) G(x)$$

Moreover, a homeomorphic solution to the equation (7.165) will be an absolute energy minimiser. We now turn to the Euler–Lagrange equations for this energy functional. They are the second order linear system of PDEs in divergence form,

$$\text{Div}[Df(x)G^{-1}(x)] = 0 \tag{7.166}$$

Among the stationary points for the energy functional are the solutions to this equation. For example if $G(x) = \mathbf{I}$, the stationary points are simply pairs $f = (u, v)$ of harmonic functions, while a couple of harmonic conjugate functions gives rise to a holomorphic function which solves the first order system of Cauchy–Riemann equations. Conformal mappings are viewed therefore, as absolute minimisers of the Dirichlet integral

$$\mathcal{E}[f] = \int_\Omega |Df(x)|^2 \, dx \tag{7.167}$$

subject to various constraints, one of which is described above. Indeed the original proof of the Riemann mapping theorem relies on this point of view. These observations fall into a general heuristic which we want to emphasize:

Quasiregular mappings are the absolute minimisers of their own energy functional and are therefore governed by the Beltrami system of first order PDEs. On the other hand, the local minima of those functionals solve the second order equation obtained by differentiating the Beltrami system.

What we mean here by "their own" energy functional, is simply the functional defined by $G = G_f$ the distortion tensor of f. We shall see that this heuristic is not simply confined to the plane. In many ways the higher dimensional theory is closely connected with the study of conformally invariant variational integrals of the form

$$\mathcal{E}[f] = \int_\Omega [\text{Trace } DfG^{-1}D^t f]^{n/2}$$

This and other related integrals represent the energy of deformations (or mappings). The absolute minimizers of which solve a first order Beltrami type system.

Our results establish the existence and regularity of minimizers in cases where the ellipticity bounds on the matrix G are significantly relaxed.

7.12. Compactness

In this section we want to present the most general convergence theorem we know in the plane. For mappings in the various classes we have considered in the previous sections, we have given modulus of continuity estimates. Together with the Arzela-Ascoli theorem, these estimates show that such families of mappings are "normal" families. That is every sequence admits a locally uniformly convergent subsequence. The modulus of continuity estimates given at Theorem 4.4 and the modulus of continuity estimate for the inverse given in §10 imply that the limit of normalized homeomorphisms is an injection. Using related ideas Tukia [**107**] showed the following theorem,

THEOREM 7.19. *Let $\Omega \subset \mathbb{C}$ be a bounded domain and $\{f_j\}_{j=1}^\infty$ be a sequence of homeomorphic mappings $f_j : \Omega \to \mathbb{C}$ such that for some $p > 0$*

$$\|\exp[K(z, f_j)]\|_{L^p(\Omega)} \leq M \tag{7.168}$$

and that

(7.169) $$f_j(a) = a, \ f_j(b) = b, \ \text{and} \ f_j(c) = c$$

for 3 distinct points in Ω. Then there is a subsequence f_{j_k} converging locally uniformly to a homeomorphism f with $\exp(K(z,f)) \in L^q(\Omega)$ for some $q > 0$.

Of course the normalization condition (7.169) can be weakened to simply assuming that f_j converge at three distinct points, to three distinct values. Also Potyemkin and Ryazanov have shown that one must take $q < p$ in general, see [**94**].

It follows from results of Brakalova and Jenkins [**26**], basically using a factorization technique, see [**58**] §10.3, that the same result is true if for some $p > 0$

$$\|K(z, f_j) \log^{-1}(e + K(z, f_j))\|_{L^p(\Omega)} \leq M$$

The important point missing here is to establish the correct dominated convergence theorem for the distortion functions. That is the lower semicontinuity properties and the regularity of the limit. Thus we want to mention the following useful addition to Tukia's theorem.

7.12.1. Biting Convergence.

We shall make use of some rather convenient terminology and machinery concerning weak convergence in L^1. This was developed by Brooks and Chacon, [**27**].

Let h and $\{h_k\}_{k=1}^\infty$ be Lebesgue measurable functions defined on a measurable set $E \subset \mathbb{C}$ and valued in a finite dimensional normed space $(\mathbb{V}, |\cdot|)$. We say that h_k converges to h in a *biting sense* if there is a sequence of measurable sets E_ν whose union is E, $E = \bigcup_{\nu=1}^\infty E_\nu$, such that $h, h_k \in L^1(E_\nu, \mathbb{V})$ for all $k, \nu = 1, 2, \ldots$ and also

(7.170) $$\lim_{k \to \infty} \int_{E_\nu} \varphi h_k = \int_{E_\nu} \varphi h,$$

whenever $\varphi \in L^\infty(E_\nu)$. Of course here one may assume that E_ν is an increasing family of sets so that the sets $E \setminus E_\nu$ (the bites from E) are decreasing. Roughly, we have the sequence $\{h_k\}$ converging weakly to h in L^1 apart from arbitrarily small bites. If $\{h_k\}_{k=1}^\infty$ converges to h in the biting sense, then we write

(7.171) $$h = \overset{*}{\lim_{k \to \infty}} h_k$$

It is immaterial for the biting limit which sequence of sets one might choose as long as the weak L^1 limit on these sets exists; different bites yield the same limit [**27**]. The following algebraic property of biting limits is straightforward.

(7.172) $$\overset{*}{\lim_{k \to \infty}}(\alpha h_k + \beta g_k) = \alpha \overset{*}{\lim_{k \to \infty}} h_k + \beta \overset{*}{\lim_{k \to \infty}} g_k,$$

of course provided the limits in the right hand side exist. Here α and β are arbitrary measurable functions, finite almost everywhere on E.

The following theorem is proved in [**58**] in all dimensions (with suitable changes made to the function $P(t)$). It generalizes an important result of [**39**].

THEOREM 7.20. *Let*

$$P(t) = \frac{t^2}{\log(e+t) \log\log(e+t)}$$

Suppose that the sequence $\{f_j\}$, $f_j : \Omega \to \mathbb{C}$, of orientation preserving mappings converges weakly in $W^{1,P}(\Omega)$ to an orientation preserving mapping $f \in W^{1,P}(\Omega, \mathbb{C})$. Suppose that

(7.173) $$K(z, f_j) \leq M_j(z), \qquad a.e. \ \Omega$$

and

(7.174) $$\overset{*}{\lim_{j \to \infty}} M_j = M(z)$$

where M is a measurable function, finite almost everywhere. Then

(7.175) $$K(z, f) \leq M(z) \qquad a.e. \ \Omega$$

7.13. Removable Singularities

In this section we shall recall a theorem analogous to the Painlevé theorem for analytic functions in the plane concerning removable singularities for bounded analytic functions. This result was established by us with K. Astala and P. Koskela [7]. We present it here simply for the sake of completeness.

There are examples to show that these results are qualitatively optimal. In particular sets of Hausdorff dimension zero are not removable for bounded mappings of exponentially integrable distortion, although they are removable for every bounded quasiregular mapping. This implies that the space of all quasiregular mappings is not dense in the space of mappings with exponentially integrable distortion. Compare this with the fact that L^∞ is not dense in BMO.

The following theorem was a first effort to look at the distortion of Hausdorff dimension under mappings with exponentially integrable distortion. It is possible that sets of Hausdorff dimension zero can be mapped to sets of large Hausdorff dimension by such mappings. Thus we must look at more subtle measures of dimension. These are Hausdorff measures with weights. We recall the notation $\mathcal{H}_h(E)$ for the Hausdorff measure of a set E with weight function $h(t)$, see [36]. The usual Hausdorff measures in $\mathbb{C}$ (corresponding to $h(t) = t^\beta$) are denoted $\mathcal{H}^\beta$ for $0 \leq \beta \leq 2$.

In what follows we set p_0 be the exponent of Theorem 7.2.

THEOREM 7.21. *Suppose that $f : \Omega \to \mathbb{C}$ is a mapping of exponentially integrable distortion with exponent $p > p_0$. If $E \subset \Omega$ is compact with $\mathcal{H}_h(E) = 0$ for the weight function $h(t) = |\log t|^{-3/2}$, $0 < t < \frac{1}{2}$, then $\mathcal{H}^1(f(E)) = 0$.*

The reader will no doubt be aware of the significance of the linear measure of $f(E)$ being 0. It allows us to apply Painlevé's theorem in a suitable setting.

THEOREM 7.22. *Let $\Omega \subset \mathbb{C}$ be a bounded domain and E a compact subset with $\mathcal{H}_h(E) = 0$ for the weight function $h(t) = |\log t|^{-3/2}$. Suppose that $f : \Omega \setminus E \to \mathbb{C}$ is a bounded mapping of exponentially integrable distortion with exponent $p \geq p_0$,*

(7.176) $$\int_\Omega e^{pK} < \infty$$

Then f extends to a mapping with exponentially integrable distortion on Ω.

COROLLARY 7.23. *If E is a compact subset of conformal capacity zero, then E is removable for bounded mappings of exponentially integrable distortion with exponent $p \geq p_0$.*

There is also the following example. Define the weight function
$$h_\alpha(t) = |\log t|^{-\alpha} \tag{7.177}$$
and let
$$\nu_\alpha(\cdot) = \mathcal{H}_{h_\alpha}(\cdot) \tag{7.178}$$
denote the weighted Hausdorff measure. We have

THEOREM 7.24. *For each $\alpha > 0$ there are sets $E_\alpha \subset \mathbb{C}$, exponents p_α and mappings $f_\alpha : \mathbf{B} \to \mathbb{C}$ of exponentially integrable distortion such that*
1. $\int e^{p_\alpha K} < \infty$
2. $\nu_\alpha(E_\alpha) < \infty$,
3. $dim(f_\alpha E_\alpha) > 1$.

Moreover, we have $p_\alpha \to \infty$ as $\alpha \to \infty$.

7.14. Final Comments

There are other aspects to the planar theory of the solution of the Beltrami equation which we have not touched upon so far and we take a moment here to point out.

For the existence theory of solutions to the Beltrami equation there are two basic approaches, the analytic and the geometric. David in his seminal paper [33] followed the analytic approach as we have largely done here. However the authors of the papers [26, 97] use a geometric approach by estimating the distortion of the moduli of certain curve families, giving in particular distortion estimates for annular rings, these are subsequently used to produce modulus of continuity estimates leading to compactness results and so forth. At present this approach is insufficiently robust to provide the detailed existence and regularity results we have achieved. However, it seems worthwhile to us that the connections between the analytic and geometric approaches be further explored. In particular a necessary and sufficient condition relating the geometric distortion of moduli and the analytic condition of exponentially integrable distortion would be particularly useful in applications. This seems to us to be a rather challenging problem, the classical equivalence for quasiconformal mappings between the analytic and geometric definitions [40] is well presented in Lehto and Virtanen's book [75].

There are other questions. Many of them natural extensions of important (and largely solved) questions from the theory of quasiconformal mappings. Suppose μ is supported in the unit disk and f is a principal solution to the associated Beltrami equation. Describe geometrically the Jordan curve $f(\partial \mathbf{B})$. What can be said about sewing maps and what is an appropriate generalization of quasisymmetry ? What of extension problems and so forth ?

Indeed, one can more or less leaf through Ahlfors' lovely little book [3] asking questions in this more general setting. Of course, the most important thing is to find substantive applications of the ideas and technology developed.

CHAPTER 8

Some Technical Results

8.1. The Divergence Condition

This section is devoted to showing that, modulo some minor technical proviso, the divergence condition at (4.27) is also sufficient to achieve the lower bound at (4.26). Precisely we have

THEOREM 8.1. *Let P be a C^∞ Orlicz function satisfying the divergence condition*

$$(8.1) \qquad \int_1^\infty P(t)\frac{dt}{t^3} = \infty.$$

Suppose that in addition for sufficiently large T and all $t > T$ we have

$$(8.2) \qquad t^{-2}(\log t)\, P(t) \geq \frac{1}{\log t}$$

and that either

- *the left hand side is bounded below by a positive constant, for instance the case $\varphi(s) = \frac{1}{s}$, or*
- *the left hand side eventually decreases monotonically to 0.*

Then there is a decreasing function $\varphi \in C^1(0,1]$ such that $\lim_{s\to 0} \varphi(s) = \infty$ and if we set

$$(8.3) \qquad \Phi(t) = \sup_{0<\epsilon<1} \frac{-1}{\varphi(\epsilon)} \int_\epsilon^1 st^{2-s} d\varphi(s)$$

as at (4.25) we have the lower bound $P(t) \geq c\, \Phi(t)$ for some positive constant c and all $t \geq 1$. In particular we have the bound as at (4.26)

$$(8.4) \qquad \varphi(\epsilon)P(t) \geq -c\int_\epsilon^1 st^{2-s} d\varphi(s)$$

for all $0 < \epsilon < 1$ and all $t \geq 1$.

Note that the bound at (8.2) involves an insignificant loss of generality as the function $t^2 \log^{-2}(1+t)$ grows too slowly for the integral at (8.1) to converge. In practice all the Orlicz functions that have any role to play in the integrability theory of Jacobians enjoy the property (8.2). For example the iterated logarithmic scales as at (3.7)

$$P(t) = \frac{t^2}{\log_1(1+t)\,\log_2(e+t)\,\log_3(e^e+t)\cdots \log_n(e^{e^{\cdot^{\cdot^{\cdot}}}}+t)}$$

fall nicely into this category.

Proof. Let us set
$$Q(t) = t^{-2} P(t), \qquad t \geq T.$$
It clearly suffices to prove (8.4) for $t \geq T > 1$, the case $1 \leq t \leq T$ being obvious. For the sake of simplicity we shall assume the normalization that
$$P(T) = T^2$$
which involves no loss of generality as we may simply scale P to achieve it.

Thus $Q(T) = 1$ and

(8.5) $$\int_T^\infty Q(t) \frac{dt}{t} = \infty$$

The case that $Q(t) \log t \geq c > 0$ is covered by simply setting $\varphi(s) = \frac{1}{s}$, but of course this is the uninteresting case. We are therefore left with the case that the function $Q(t) \log t$, monotonically decreases to 0 for $t \geq T$. We also have our technical assumption

(8.6) $$Q(t) \geq \log^{-2} t$$

for all $t \geq T$.

We begin by looking for φ in the form

(8.7) $$\varphi(s) = \exp\left(\int_s^1 \frac{d\nu(\tau)}{\tau} \right), \qquad 0 < s \leq 1$$

for some bounded and nondecreasing function ν. To ensure that we have $\lim_{s \to 0} \varphi(s) = \infty$ we need that

(8.8) $$\int_0^1 \frac{d\nu(s)}{s} = \infty$$

On substituting (8.7) into (8.4) we arrive at the inequality

(8.9) $$\frac{c}{\varphi(\epsilon)} \int_\epsilon^1 \frac{\varphi(s) d\nu(s)}{t^s} \leq Q(t)$$

for all $0 < \epsilon \leq 1$ and all $t \geq T$. As φ as given at (8.7) will be decreasing we will always have $\varphi(s) \leq \varphi(\epsilon)$ for $\epsilon \leq s \leq 1$. Thus the problem now reduces to showing the inequivalent fact that

(8.10) $$\int_0^1 t^{-s} d\nu(s) \leq C\, Q(t)$$

for some positive constant C and all sufficiently large values of t.

At this point it is convenient to introduce another decreasing function

(8.11) $$\Theta(x) = Q(e^x), \qquad \log T \leq x \leq \infty$$

Then the inequality at (8.10) translates into an inequality for the Laplace transform, namely

(8.12) $$\int_0^1 e^{-sx} d\nu(s) \leq C\Theta(x)$$

at least for $x \geq \log T$. Notice that

(8.13) $$\int_{\log T}^\infty \Theta(x)\, dx = \int_{\log T}^\infty Q(e^x)\, dx = \int_T^\infty Q(t) \frac{dt}{t} = \infty$$

8.1. THE DIVERGENCE CONDITION

It is both rewarding and illuminating to digress a little at this point. We recall a theorem of Bernstein concerning the Laplace transform, see [**111**] and also [**44**]. A function $\Theta = \Theta(x)$ is said to be completely monotonic in the interval $[0, \infty)$ if

(8.14) $$(-1)^k \frac{d^k \Theta}{dx^k} \geq 0, \quad k = 0, 1, 2, \ldots$$

Bernstein's Theorem asserts that if Θ is a completely monotonic function, then there exists a positive bounded and nondecreasing function $\nu = \nu(s)$, $0 \leq s < \infty$, such that as a Stieltjes integral we have

(8.15) $$\Theta(x) = \int_0^\infty e^{-sx} d\nu(s)$$

Now we would have established our desired lower bound (8.4) if our function

$$\Theta(x) = Q(e^x) = e^{-2x} P(e^x)$$

was completely monotonic. Indeed in that case we would have

$$\Theta(x) = \int_0^\infty e^{-sx} d\nu(s) = \int_0^1 e^{-sx} d\nu(s) + \int_1^\infty e^{-sx} d\nu(s)$$
$$\leq \int_0^1 e^{-sx} d\nu(s) + \nu(\infty) e^{-x}$$

and hence

$$\int_0^1 \frac{d\nu(s)}{s} = \int_0^1 \left(\int_0^\infty e^{-sx} dx \right) d\nu(s)$$
$$\geq \int_0^\infty \Theta(x) dx - \nu(\infty) = \infty$$

which ensures the condition (8.8).

Unfortunately the notion of a completely monotonic function is insufficiently general to be of much use in our setting, relying as it does on all the derivatives of Θ. The reader may care to notice that in fact complete monotonicity is also necessary for a function to be represented as an integral of the form (8.15). However, we only need that Θ is bounded below by such an integral. This, as might be expected, will not depend on derivatives of Θ.

Returning now to the proof, we define $d\nu(s)$ explicitly by means of the inverse function to Θ. Let us denote this inverse by $h : (0, 1] \to [\log T, \infty)$. We set

(8.16) $$d\nu(s) = s^3 h(s^3) \, ds$$

By the interpretation of the integral as the area under a graph, we see immediately that

(8.17) $$\int_0^1 h(\epsilon) \, d\epsilon = \int_{\log T}^\infty \Theta(x) \, dx = \infty$$

Thus,

$$\int_0^1 \frac{d\nu(s)}{s} = \int_0^1 s^2 h(s^3) \, ds = \frac{1}{3} \int_0^1 h(\epsilon) \, d\epsilon = \infty$$

as required. To prove (8.12) we write $x = h(\epsilon)$ and find (8.12) reduces to establishing that

(8.18) $$\int_0^1 e^{-sh(\epsilon)} d\nu(s) \leq C \epsilon$$

for all $0 \leq \epsilon < 1$. We break the integral into three parts:

Part 1. This first estimate is easy,
$$\int_0^\epsilon e^{-sh(\epsilon)} d\nu(s) \leq \int_0^\epsilon s^3 h(s^3) ds \leq \epsilon\, h(1) = \epsilon \log T$$
Here we have used the fact that the function $\epsilon \mapsto \epsilon h(\epsilon)$ is increasing in $0 < \epsilon \leq 1$. This observation simply means that the function $x \mapsto x\Theta(x) = (\log e^x)Q(e^x)$ is monotonically decreasing for $x \geq \log T$, which in turn is simply the monotonicity of $t \mapsto Q(t)\log t = t^{-2}P(t)\log t$ for $t \geq T$ which formed part of our hypotheses for Theorem 8.1.

Part 2. Next we have
$$\int_\epsilon^{\epsilon^{1/3}} e^{-sh(\epsilon)} d\nu(s) = \int_\epsilon^{\epsilon^{1/3}} \frac{s^3 h(s^3)}{e^{sh(\epsilon)}} ds$$
$$\leq \epsilon h(\epsilon) \int_\epsilon^{\epsilon^{1/3}} e^{-sh(\epsilon)} ds$$
$$\leq \frac{\epsilon h(\epsilon)}{h(\epsilon)} e^{-\epsilon h(\epsilon)} \leq \epsilon$$
This time we have crucially used the monotonicity property of $\epsilon \mapsto \epsilon h(\epsilon)$ as noted above.

Part 3. We have for all $0 < \epsilon \leq 1$
$$\int_{\epsilon^{1/3}}^1 e^{-sh(\epsilon)} d\nu(s) \leq e^{\epsilon^{1/3} h(\epsilon)} \int_0^1 s^3 h(s^3)\, ds$$
$$\leq h(1) e^{-\epsilon^{1/3} h(\epsilon)}$$
$$\leq (e^6 \log T)\epsilon$$

The inequalities found here are argued as follows. By (8.6) we see that $x^2 Q(e^x) \geq 1$, for $x \geq \log T$. Hence $x^2 \Theta(x) \geq 1$, which is the same thing as $(h(\epsilon))^2 \epsilon \geq 1$ upon the substitution $x = h(\epsilon)$. Finally for all $0 < \epsilon \leq 1$, we have
$$\epsilon^{1/3} h(\epsilon) \geq \frac{1}{\epsilon^6} \geq \log(1/\epsilon)) - 6$$
as desired.

This completes the proof of Theorem 8.1. $\square$

8.2. Integration by Parts

In this secction we wish to provide an abbreviated foundation of the sublinear integrability theory of the Jacobian determinants. Our presentation is primarily based on the inequality appearing in the following theorem.

THEOREM 8.2. *Let $F : \mathbb{C} \to \mathbb{C}$ be a mapping belonging to the Sobolev space $W^{1,2-s}(\mathbb{C})$, $0 < s \leq 1$. Then*

(8.19) $$\left| \int_\mathbb{C} |DF(z)|^{-s} J(z, F)\, dz \right| \leq A\, s \int_\mathbb{C} |DF(z)|^{2-s}\, dz$$

where A is a constant independent of s.

The proof of this estimate is to be found in [**51**]. Now the lower bound at (4.26) or (8.4) above, yields the inequality

$$\left| \int_{\mathbb{C}} J(z, F) \left[\frac{-1}{\varphi(\epsilon)} \int_{\epsilon}^{1} |DF(z)|^{-s} \, d\varphi(s) \right] dz \right|$$
$$\leq A \int_{\mathbb{C}} |DF(z)|^2 \left[\frac{-1}{\varphi(\epsilon)} \int_{\epsilon}^{1} s|DF(z)|^{-s} \, d\varphi(s) \right] dz$$
$$\leq A \int_{|DF| \leq 1} |DF(z)| \, dz + \frac{A}{c} \int_{|DF| > 1} P(|DF(z)|) \, dz < \infty$$

whenever F is compactly supported with the latter integral converging. In this estimate we want to pass to the limit as $\epsilon \to 0$. By the Lebesgue Dominated Convergence Theorem and L'Hôpital's rule we find that

$$\lim_{\epsilon \to 0} \int_{\mathbb{C}} |DF(z)|^2 \left[\frac{-1}{\varphi(\epsilon)} \int_{\epsilon}^{1} s|DF(z)|^{-s} \, d\varphi(s) \right] dz = 0$$

and hence

(8.20) $$\lim_{\epsilon \to 0} \int_{\mathbb{C}} J(z, F) \left[\frac{-1}{\varphi(\epsilon)} \int_{\epsilon}^{1} |DF(z)|^{-s} \, d\varphi(s) \right] dz = 0$$

The next goal we seek is to pass the limit under the integral sign and deduce that $J(z, F)$ is integrable with L^1 mean equal to 0. Unfortunately the Lebesgue Dominated Convergence Theorem does not apply in this case. We must turn to the Fatou Lemma. This will require that the integrand be non-negative, at least its principal part (modulo terms which are uniformly majorized by an integrable function).

To facilitate this we restrict our attention to an arbitrary *orientation preserving* mapping $f = f^1 + if^2 : \Omega \to \mathbb{C}$ which lies in the Orlicz–Sobolev space $W^{1,P}_{loc}(\Omega)$. Here, and in what follows we shall assume that the Orlicz function $P = P(t)$ satisfies the divergence criteria at (8.1) together with the monotonicity condition (8.2) so that we may use Theorem 8.1. With these restrictions we can now establish the following important formula for integration by parts, the applications of which the reader will no doubt be aware of.

THEOREM 8.3. *With f and P satisfying the hypotheses above we have that the Jacobian determinant $J(z, f) \geq 0$ is locally integrable and enjoys the formula of integration by parts:*

(8.21) $$\int_{\Omega} \eta(z) \, df^1 \wedge df^2 = - \int_{\Omega} f^1(z) d\eta \wedge df^2$$

for all Lipshitzian test functions η with compact support in Ω.

Proof. First notice that the integral in the right hand side converges as

$$|Df| \in \bigcap_{1 \leq p < 2} L^p_{loc}(\Omega), \qquad |f| \in \bigcap_{1 < r < \infty} L^r_{loc}(\Omega)$$

We need only show that the 2-form $d(\eta f^1) \wedge df^2$ is integrable and the integral of this form vanishes. We may of course assume that $\eta \geq 0$, for otherwise we decompose η into its positive and negative parts and apply (8.21) to each of these.

For computational convenience we introduce another real–valued test function $\alpha \in C_0^\infty(\Omega)$ with $\alpha(z) \equiv 1$ on the support of η. Then the mapping
$$F(z) = \eta(z)f^1 + i\alpha(z)f^2(z)$$
has compact support and belongs to $W^{1,2-s}(\mathbb{C})$ for every $0 < s \leq 1$.

In applying (8.20) we first split the Jacobian determinant as
$$\begin{aligned} J(z,F)\,dz &= d(\eta f^1) \wedge d(\alpha f^2) = d(\eta f^1) \wedge df^2 \\ &= \eta df^1 \wedge df^2 + f^1 d\eta \wedge df^2 \end{aligned}$$

Hence

(8.22)
$$\lim_{\epsilon \to 0} \left\{ \int_\Omega \eta(z) J(z,f) \left[\frac{-1}{\varphi(\epsilon)} \int_\epsilon^1 \frac{d\varphi(s)}{|DF(z)|^s} \right] dz \right.$$
$$\left. + \int_\Omega f^1(z) d\eta \wedge df^2 \left[\frac{-1}{\varphi(\epsilon)} \int_\epsilon^1 \frac{d\varphi(s)}{|DF(z)|^s} \right] dz \right\} = 0$$

First, passing the limit under the integral sign in the second term poses no real problem as an application of the Dominated Convergence Theorem. In view of the pointwise estimate $|DF(z)| \geq |df^2(z)|$ on the support of η we have the uniform bound on the integrand

$$\left| f^1(z) d\eta \wedge df^2 \left[\frac{-1}{\varphi(\epsilon)} \int_\epsilon^1 \frac{d\varphi(s)}{|DF(z)|^s} \right] \right|$$
$$\leq \frac{-1}{\varphi(\epsilon)} \int_\epsilon^1 |f^1(z)|\,|d\eta(z)|\,|df^2(z)|^{1-s}\,d\varphi(s)$$
$$\leq |f(z)|\,|d\eta(z)|(|Df(z)| + 1) \in L^1(\Omega)$$

Hence, once again employing L'Hôpitals rule, we may write
$$\lim_{\epsilon \to 0} \int_\Omega \eta(z) J(z,f) \left[\frac{-1}{\varphi(\epsilon)} \int_\epsilon^1 \frac{d\varphi(s)}{|DF(z)|^s} \right] dz = -\int_\Omega f^1(z)\,d\eta \wedge df^2$$

It is at this point that we really need the assumption that $J(z,f) \geq 0$. We find from the Fatou Lemma that $\eta(z)\,J(z,f) \in L^1(\Omega)$. Indeed
$$\int_\Omega \eta(z) J(z,f) dz \leq \lim_{\epsilon \to 0} \int_\Omega \eta(z) J(z,f) \left[\frac{-1}{\varphi(\epsilon)} \int_\epsilon^1 \frac{d\varphi(s)}{|DF(z)|^s} \right] dz$$

again by L'Hôpitals rule. Now knowing this, the Dominated Convergence Theorem applies because of the pointwise estimate
$$\frac{\eta(z) J(z,f)}{|DF(z)|^s} \leq \eta(z) J(z,f) + \eta(z)|Df(z)| \in L^1(\Omega)$$

for all $0 < s \leq 1$. This completes the proof of the theorem. $\square$

8.3. Higher Integrability

As a matter of fact the Jacobian determinant of an orientation preserving mapping $f \in W^{1,P}_{loc}(\Omega)$ enjoys a higher degree of integrability than mere L^1–integrability. The idea goes back to the work of Müller [90] and can be further traced back to Gehring [38], see also the survey article [53]. The result as we present it, is principally based on maximal inequalities in Orlicz spaces, see [20, 44].

8.3. HIGHER INTEGRABILITY

We recall that the Hardy–Littlewood maximal operator defined on a domain $\Omega \subset \mathbb{C}$ is given by the rule

$$(8.23) \qquad (\mathcal{M}_\Omega h)(x) = \sup\left\{\frac{1}{|Q|}\int_Q |h|\right\}$$

with the supremum running over all cubes Q in Ω which contain x. There is no loss in generality in our discussion if we assume the Orlicz function Φ satisfies

$$\int_0^1 \Phi(s)\frac{ds}{s^2} < \infty$$

because in order to achieve this we need only modify Φ near 0. Given such a Φ there is a corresponding Hardy–Littlewood conjugate function

$$(8.24) \qquad \Psi(t) = \Phi(t) + t\int_0^t \Phi(s)\frac{ds}{s^2} \geq \Phi(t)$$

The reader should compare this with (5.12) in §6.2. It is worth pointing out here that Ψ grows at least linearly; and it is actually convex if, for instance, $t\Phi'(t)$ increases. We now recall the Maximal Theorem in the setting of Orlicz spaces, see [13, 44] for proofs.

THEOREM 8.4. *Let $Q \subset \mathbb{C}$ be a cube. For each non-negative measurable function h defined on Q we have*

$$(8.25) \qquad \frac{1}{|Q|}\int_Q \Psi(h) - \Psi\left(\frac{1}{|Q|}\int_Q h\right) \leq \frac{4}{|Q|}\int_Q \Phi(\mathcal{M}_Q h)$$

Moreover, if the function $t \mapsto t^{-p}\Phi(t)$ is increasing for some $p > 1$, then

$$(8.26) \qquad \int_Q \Phi(\mathcal{M}_Q h) \leq \frac{9p}{p-1}\int_Q \Phi(2h)$$

Now let $P = P(t)$ be an Orlicz function satisfying the divergence condition (8.1) and the monotonicity condition (8.2). We define

$$(8.27) \qquad R(t) = t\int_0^{\sqrt{t}} P(s)\frac{ds}{s^3} = 2t\int_0^t P(\sqrt{t})\frac{ds}{s^2}$$

We are now able to establish the following theorem.

THEOREM 8.5. *Let $f \in W^{1,P}(\Omega)$ be an orientation preserving mapping. Then the Jacobian determinant belongs to the Orlicz space $L^R_{loc}(\Omega)$, and we have the uniform bounds*

$$(8.28) \qquad \|J(z,f)\|_{L^R(X)} \leq C(X)\|Df\|^2_{L^P(\Omega)}$$

for every relatively compact $X \subset \Omega$.

This theorem is sharp in the following sense: The conclusion at (8.28) fails if $R(t)$ is replaced by any other Orlicz function $\lambda(t)R(t)$ with $\lim_{t\to\infty}\lambda(t) = \infty$.

Sketch of Proof. We only give the main points of the calculation. We may assume that Ω is a cube and that X is a concentric cube of half the size. The formula for integration by parts (8.21) yields

$$\int_\Omega \eta(z)J(z,f)\,dz \leq \int_\Omega |\nabla\eta(z)|\,|f(z)|\,|Df(z)|\,dz$$

We apply the Poincaré–Sobolev inequality to arrive at the local estimates

$$\frac{1}{|Q|}\int_Q J(z,f)\,dz \le C\left(\frac{1}{|2Q|}\int_{2Q}|Df(z)|^{4/3}\,dz\right)^{3/2}$$

for every cube Q whose double $2Q$ is contained in Ω. Setting $J = J(z,f)$ and $H = |Df(z)|^{4/3}$, these inequalities translate to the pointwise estimate of the maximal functions

$$(\mathcal{M}_X J)(z) \le C\,(\mathcal{M}_\Omega H)^{3/2}(z)$$

for all $z \in X$. By the maximal inequality of Theorem 8.4 it follows that

$$\|J\|_{L^R(X)} \le C\|\mathcal{M}_X J\|_{L^\Phi(X)}$$

where

$$\Phi(t) = P(\sqrt{t}), \qquad R(t) = t\int_0^t \Phi(s)\,\frac{ds}{s^2}$$

Hence

$$\begin{aligned}
\|J\|_{L^R(X)} &\le C\,\|(\mathcal{M}_\Omega H)^{3/2}\|_{L^\Phi(X)} \\
&\le C\,\|(\mathcal{M}_\Omega H)^{3/2}\|_{L^\Phi(\Omega)} \\
&\le C\,\|H^{3/2}\|_{L^\Phi(\Omega)} = C\|\,|Df|^2\,\|_{L^\Phi(\Omega)} \\
&\le C\,\|Df\|^2_{L^P(\Omega)}
\end{aligned}$$

as desired. □

It is worthwhile to present a few examples here so that the reader can see what sorts of Orlicz functions are related by Theorem 8.5. We have

- $P(t) = t^2 \log^\alpha(1+t)$ and $R(t) \approx t\log^{1+\alpha}(1+t)$ for all real $\alpha \ne -1$.
- $P(t) = t^2 \log^{-1}(1+t)$ and $R(t) \approx t\log\log(e+t)$.
- And more generally, the iterated logarithms functions give

$$P(t) = \frac{t^2}{\log_1(1+t)\,\log_2(e+t)\,\log_3(e^e+t)\cdots\log_n(e^{e^{\cdot^{\cdot^{\cdot}}}}+t)}$$

with

$$R(t) \approx t\log_{n+1}(1+t)$$

Notice that formula (8.27) for R never gives $R(t) \approx c\,t$, though we can come quite close to it. The gain (or jump) in the degree of integrability of the Jacobian when compared with that of $|Df|^2$ can be measured by the ratio

$$(8.29)\qquad L(t) = \frac{R(t)}{P(\sqrt{t})} \approx \log_1(1+t)\,\log_2(e+t)\cdots\log_{n+1}(e^{e^{\cdot^{\cdot^{\cdot}}}}+t)$$

Our example suggests the following heuristic principle:

The greatest gains in regularity are to be found closest to the L^1-summability of the Jacobian determinant.

In any case the limitation for the gain (or jump) in regularity is imposed by the condition

$$\int_0^\infty \frac{dt}{tL(t)} = \infty$$

Bibliography

[1] L.V. Ahlfors, *Zur theorie der Überlagerungsflächen*, Acta Math., **65**, (1935).
[2] L.V. Ahlfors, *Commentary on: "Zur theorie der Überlagerungsflächen" (1935)*, Fields Medallists' lectures, 8–9, World Sci. Ser. 20th Century Math., **5**, World Sci. Publishing, River Edge, NJ, 1997.
[3] L.V. Ahlfors, *Lectures on quasiconformal mappings*, Van Nostrand, Princeton 1966; Reprinted by Wadsworth Inc. Belmont, 1987.
[4] L.V. Ahlfors, *On quasiconformal mappings*, J. Anal. Math., **3**, (1953/54), 1–58.
[5] L.V. Ahlfors and A. Beurling, *Conformal invariants and function theoretic null sets*, Acta Math., **83**, (1950), 101–129.
[6] K. Astala, *Area distortion of quasiconformal mappings*, Acta Math., **173**, (1994), 37–60.
[7] K. Astala, T. Iwaniec, P. Koskela and G.J. Martin, *Mappings with BMO bounded Distortion*, Math. Annalen, **317**, (2000), 703-726.
[8] K. Astala, T. Iwaniec and G.J. Martin, *Elliptic Equations and Quasiconformal Mappings in the Plane*, Monograph, to appear.
[9] K. Astala, T. Iwaniec, G.J. Martin and J. Onninen, *Extremal mappings of finite distortion*, Proc. London Math. Soc., **91**, (2005), 655-702.
[10] K. Astala, T. Iwaniec, and E. Saksman, *Beltrami operators in the plane*, Duke Math. J., **107**, (2001), 27–56.
[11] K. Astala and G.J. Martin, *Holomorphic Motions*, Papers on Analysis, A volume dedicated to Olli Martio on the occasion of his 60th Birthday, Report Univ. Jyvskyl, **83**, (2001), 27-40.
[12] J. Ball, *Convexity conditions and existence theorems in nonlinear elasticity*, Arch. Rat. Mech. Anal., **63**, (1977), 337-403.
[13] R. Bagby and D. Parson, *Orlicz spaces and rearranged maximal functions*, Math. Nachr., **132**, (1987), 15–27.
[14] R. Banuelos and G. Wang, *Sharp inequalities for martingales with applications to the Beurling-Ahlfors and Riesz transforms*, Duke Math. J., **80**, (1995), 575–600.
[15] E.T. Bell, *The development of Mathematics*, 2nd Ed., McGraw-Hill, 1945.
[16] E. Beltrami, *Saggio di interpretazione della geometria non euclidea*, Giornale di Mathematica, **6**, 1867.
[17] L. Bers, *On a theorem of Mori and the definition of quasiconformality*, Trans. Amer. Math. Soc., **84**, (1957), 78–84.
[18] L. Bers, *Mathematical aspects of subsonic and transonic gas dynamics*, Surveys in Applied Mathematics, **3**, John Wiley and Sons, Inc., New York; Chapman & Hall, Ltd., London 1958.
[19] L. Bers, *Uniformisation by Beltrami equations*, Comm. Pure Appl. Math., **14**, (1961), 215–228.
[20] H. Brezis, N. Fusco and C. Sbordone, *Integrability for the Jacobian of orientation preserving mappings*, J. Funct. Anal., **115**, (1993), 425–431.
[21] B. Bojarski, *Homeomorphic solutions of Beltrami systems*, Dokl. Akad. Nauk. SSSR, **102**, (1955), 661–664.
[22] B. Bojarski, *Generalised solutions of an elliptic system of the first order with discontinuous coefficients*, Mat. Sbornik, **43**, (1957), 451–503.
[23] B. Bojarski, *Subsonic flow of compressible fluid*, Arch. Mech. Stos., **18**, (1966), 497–520.
[24] B. Bojarski and T. Iwaniec, *Quasiconformal mappings and nonlinear elliptic equations in two variables, I & II*, Bull. Polish. Acad. Sci., XXII, (5), (1974), 473–478.
[25] A. Bonami, T. Iwaniec, P. Jones and M. Zinsmeister, *On the product of functions in BMO and H^1*, to appear, Ann. Inst. Fourier.

[26] M.A. Brakalova and J.A. Jenkins, *On solutions of the Beltrami equation*, J. Anal. Math., **76**, (1998), 67–92.

[27] J.K. Brooks and R.V. Chacon, *Convergence theorems in the theory of diffusions*, Measure theory and its applications, 79–93, Lecture Notes in Math., **1033**, Springer, Berlin-New York, 1983.

[28] R. Coifman and G. Weiss, *Analyse Harmonique Non-commutative sur Certain Espaces Homogènes*, Lecture Notes in Math., **242**, Springer-Verlag, 1971.

[29] D-C. Chang, S.G. Krantz and E.M. Stein, H^p *theory on a smooth domain in* $\mathbb{R}^N$ *and elliptic boundary value problems*, J. Funct. Anal., **114**, (1993), 286–347.

[30] R.R. Coifman, P.L Lions, Y. Meyer and S. Semmes, *Compensated compactness and Hardy spaces*, J. Math. Pures Appl., **72**, (1993), 247–286.

[31] R.R. Coifman and R. Rochberg, *Another characterization of BMO*, Proc. Amer. Math. Soc., **79**, (1980), 249–254.

[32] R.R. Coifman, R. Rochberg and G. Weiss, *Factorization theorems for Hardy spaces in several variables*, Ann. Math., **103**, (1978), 569–645.

[33] G. David, *Solutions de l'equation de Beltrami avec* $\|\mu\|_\infty = 1$, Ann. Acad. Sci. Fenn. Ser. AI Math., **13**, (1988), 25–70.

[34] A. Douday, *Prolongement de mouvements holomorphes (d'après Słodkowski et autres)*, Séminaire Bourbaki, Vol. 1993/94. Astérisque No. 227, (1995), Exp. No. 775, **3**, 7–20.

[35] C. Fefferman, *Characterisations of bounded mean oscillation*, Bull. Amer. Math. Soc., **77**, (1971), 587–588.

[36] J. Garnett, *Bounded analytic functions*, Academic Press, 1972.

[37] F.W. Gehring, *Rings and quasiconformal mappings in space*, Trans. Amer. Math. Soc., **103**, (1962), 353–393.

[38] F.W. Gehring, *The L^p-integrability of the partial derivatives of a quasiconformal mapping*, Acta Math., **130**, (1973), 265–277.

[39] F.W. Gehring and T. Iwaniec, *The limit of mappings with finite distortion*, Ann. Acad. Sci. Fenn. Math., **24**, (1999), 253–264.

[40] F.W. Gehring and O. Lehto, *On the total differentiability of functions of a complex variable*, Ann. Acad. Sci. Fenn. A I, **272**, (1959), 9pp.

[41] D. Gilbarg and N.S. Trudinger, *Elliptic Partial Differential Equations of Second Order*, Springer-Verlag, 1983.

[42] V.M. Goldstein and S.K Vodop'yanov, *Quasiconformal mappings and spaces of functions with generalised first derivatives*, Sb. Mat. Z., **17**, (1976), 515–531.

[43] L. Greco, *A remark on the equality* $\det Df = \text{Det } Df$, Differential and Integral Equations, **6**, (1993), 1089–1100.

[44] L. Greco, T. Iwaniec and G. Moscariello, *Limits of the improved integrability of the volume forms*, Indiana Math. J., **44**, (1995), 305–339.

[45] H. Grötzsch *Über die Verzerrung bei schlichten nichtconformen Abbildungen und übereine damit zusammenhängende Erweiterung des Picardschen Satzes* er. Verh. Sächs. Akad. Wiss. Leipzig, **80**, (1928), 503–507.

[46] P. Haissinsky and L. Tan, *Convergence of pinching deformations and matings of geometrically finite polynomials*, Fund. Math., **181**, (2004), 143–188.

[47] P. Hajlasz, T. Iwaniec, J. Malý and J. Onninen, *Weakly differentiable mappings between manifolds*, to appear, Memoirs Amer. Math. Soc.

[48] J. Heinonen and P. Koskela *Sobolev mappings with integrable dilatation*, Arch. Rat. Mech. Anal., **125**, (1993), 81–97.

[49] S. Hencl and P. Koskela, *Regularity of the inverse of a planar Sobolev homeomorphism*, Arch. Ration. Mech. Anal. , **180**, (2006), 75–95.

[50] J. Hogan, C. Li, A. McIntosh and K. Zhang, *Global higher integrability of Jacobians on bounded domains*, Ann. Inst. H. Poincar Anal. Non Linaire, **17**, (2000), 193–217.

[51] T. Iwaniec, *Nonlinear Cauchy–Riemann operators in* $\mathbb{R}^n$, Trans. Amer. Math. Soc., **354**, (2002), 1961–1995.

[52] T. Iwaniec, *Quasiconformal mapping problem for general nonlinear systems of partial differential equations*, Symposia Mathematica, XVIII, (1976), 501–517.

[53] T. Iwaniec, *The Gehring Lemma*, Quasiconformal Mappings and Analysis, Ed. P. Duren, J. Heinonen, B. Osgood and B. Palka, Springer–Verlag, 1998.

[54] T. Iwaniec, P. Koskela and G.J. Martin, *Mappings with BMO-distortion*, J. Anal. Math., **88**, (2002), 337–381.
[55] T. Iwaniec, P. Koskela, G. Martin and C. Sbordone, *Mappings of exponentially integrable distortion: $L^n \log^\alpha L$-integrability*, J. London Math. Soc., **67**, (2003) 123–136.
[56] T. Iwaniec, P. Koskela and J. Onninen, *Mappings of finite distortion: Monotonicity and Continuity*, Invent. Math., **144**, (2001), 507–531.
[57] T. Iwaniec and G.J. Martin *Quasiregular mappings in even dimensions*, Acta Math., **170**, (1993), 29–81.
[58] T. Iwaniec and G.J. Martin *Geometric function theory and non-linear analysis*, Oxford Mathematical Monographs. The Clarendon Press, Oxford University Press, New York, 2001.
[59] T. Iwaniec and G.J. Martin *Squeezing the Sierpiński sponge*, Studia Math., **149**, (2002), 133–145.
[60] T. Iwaniec and G.J. Martin *Quasiconformal mappings and capacity*, Indiana Univ. Math. J., **40**, (1991), 101–122.
[61] T. Iwaniec and G.J. Martin *The geometric analysis of deformations of finite distortion: future directions and problems*, Future trends in geometric function theory, 119–142, Rep. Univ. Jyvskyl Dep. Math. Stat., 92, Univ. Jyvskyl, Jyvskyl, 2003.
[62] T. Iwaniec, L. Migliaccio, G. Moscariello and A. Passarelli, *A priori estimates for nonlinear elliptic complexes*, Advances in Differential Equations, **8**, (2003), 513–546.
[63] T. Iwaniec and C. Sbordone, *On the integrability of the Jacobian under minimal hypotheses*, Arch. Rational Mech. Anal., **119**, (1992), 129–143.
[64] T. Iwaniec and V. Šverák, *On mappings with integrable dilatation*, Proc. Amer. Math. Soc., **118**, (1993), 181–188.
[65] T. Iwaniec and A. Verde, *A study of Jacobians in Hardy-Orlicz spaces*, Proc. Roy. Soc. Edinburgh Sect., A, **129**, (1999), 539–570.
[66] S. Janson, *Generalizations of Lipschitz spaces and an application to Hardy spaces and bounded mean oscillation*, Duke Math. J., **47**, (1980), 959–982.
[67] F. John and L. Nirenberg, *On functions of bounded mean oscillation*, Comm. Pure Appl. Math., **14**, (1961), 415–426.
[68] J. Kauhanen, P. Koskela and J. Malý, *On functions with a derivative in a Lorentz space*, Manuscript Math., **100**, (1999), 87–101.
[69] J. Kauhanen, P. Koskela and J. Malý, *Mappings of finite distortion: discreteness and openness*, Arch. Ration. Mech. Anal., **160**, (2001), 135–151.
[70] J. Kauhanen, P. Koskela and J. Malý, *Mappings of finite distortion: condition N.*, Michigan Math. J., **49**, (2001), 169–181.
[71] M.A. Lavrentiev, *Sur une critère differentiel des transformations homéomorphes des domains à trois dimensions*, Dokl. Acad. Nauk. SSSR, **20**, (1938), 241–242.
[72] O. Lehto, *Homeomorphisms with a given dilatation*, Proc. 15th Scandinavian Conference, Oslo 1968, Lecture Notes in Math., **118**, Springer–Verlag, (1970), 58–73.
[73] O. Lehto, *Remarks on generalized Beltrami equations and conformal mappings*, Proc. Romainian–Finnish seminar on Teichmüller spaces and quasiconformal mappings, Romania 1969, Publishing house of the Academy of the Socialist Republic of Romania, Bucharest, (1971), 203–214.
[74] O. Lehto, *Univalent functions and Teichmüller spaces*, Springer–Verlag, 1987.
[75] O. Lehto and K. Virtanen, *Quasiconformal mappings in the plane*, Springer–Verlag, 1971.
[76] Z. Lou and A. McIntosh, *Hardy spaces of exact forms on Lipschitz domains in $\mathbb{R}^N$*. Indiana Univ. Math. J., **53**, (2004), 583–611.
[77] J. Malý and O. Martio, *Lusin's condition (N) and mappings of the class $W^{1,n}$*, J. Reine Angew. Math., **458**, (1995), 19–36.
[78] R. Mañé, P. Sad and D. Sullivan, *On the dynamics of rational maps*, Ann. Sci. École Norm. Sup., **16**, (1983), 193–217.
[79] J. Manfredi, *Weakly monotone functions*, J. Geometric Analysis, **3**, (1994), 393-402.
[80] J. Manfredi and E. Villamor, *Mappings with integrable dilatation in higher dimensions*, Bull. Amer. Math. Soc., **32**, (1995), 235–240.
[81] J. Manfredi and E. Villamor, *An extension of Reshetnyak's Theorem*, Indiana Math. J., **47**, (1998), 1131–1145.
[82] B. Maskit, *Kleinian Groups*, Springer–Verlag, 1987.

[83] C.T. McMullen, *Renormalization and 3-manifolds which fiber over the circle*, Annals of Mathematics Studies, **142**, Princeton University Press, Princeton, NJ, 1996.

[84] C.T. McMullen, *Complex dynamics and renormalization*. Annals of Mathematics Studies, **135**, Princeton University Press, Princeton, NJ, 1994.

[85] A. Miyachi, H^p *spaces over open subsets of* $\mathbb{R}^n$. Studia Math., **95**, (1990), 205–228.

[86] L. Migliaccio and G. Moscariello, *Higher integrability of div-curl products*, Richerche di Matematica, **49**, (2000)

[87] C.B. Morrey, *On the solutions of quasi-linear elliptic partial differential equations*, Trans. Amer. Math. Soc., **43**, (1938), 126–166.

[88] C.B. Morrey, *Multiple integrals in the calculus of variations*, Die Grundlehren der mathematischen Wissenschaften, Band 130 Springer-Verlag New York, Inc., New York 1966.

[89] G. Moscariello, *On the integrability of the Jacobian in Orlicz spaces*, Math. Japonica, **40**, (1994), 323–329.

[90] S. Müller, *A surprising higher integrability property of mappings with positive determinant*, Bull. Amer. Math. Soc., **21**, (1989), 245–248.

[91] S. Müller, T. Qi and B. Yan, *On a new class of elastic deformations not allowing for cavitation*, Ann. Inst. H. Poincaré Anal. Non Linaire, **11**, (1994), 217–243.

[92] S. Müller and S. Spector, *An existence theory for non-linear elasticity that allows for cavitation*, Arch. Rational Mech. Anal., **131**, (1995), 1-66.

[93] A. Pfluger, *Une propriété métrique de la représentation quasi conforme*, C. R. Acad. Sci., Paris, 226, (1948), 623–625.

[94] V. Potyemkin, and V. Ryazanov, *On the noncompactness of David classes*, Ann. Acad. Sci. Fenn. Math., **23**, (1998), 191–204.

[95] M.M Rao and Z.D. Ren, *Theory of Orlicz Spaces*, Pure and Applied Math. **146**, New York, 1991.

[96] M. Reimann, *Functions of bounded mean oscillation and quasiconformal mappings*, Comment. Math. Helv., **49**, (1974), 260–276.

[97] V. Ryazanov, U. Srebro and E. Yakubov, *BMO-quasiconformal mappings*, J. Anal. Math., **83**, (2001), 1–20.

[98] V. Ryazanov, U. Srebro and E. Yakubov, *On ring solutions of Beltrami equations*, J. Anal. Math., **96**, (2005), 117–150.

[99] M. Shishikura, *On the quasiconformal surgery of rational functions.*, Ann. Sci. École Norm. Sup., **20**, (1987), 1–29.

[100] Z. Slodkowski, *Holomorphic motions and polynomial hulls*, Proc. Amer. Math. Soc., **111**, (1991), 347–355.

[101] U. Srebro and E. Yakubov, *Branched folded maps and alternating Beltrami equations*, J. Anal. Math., **70**, (1996), 65–90.

[102] E.M Stein, *Singular integrals and differentiability properties of functions*, Princeton University Press, 1970.

[103] E.M. Stein, *Note on the class* $L \log L$, Studia Math., **32**, (1969), 305–310.

[104] D. Sullivan, *Quasiconformal homeomorphisms and dynamics, I. Solution of the Fatou-Julia problem on wandering domains*. Ann. Math., **122**, (1985), 401–418.

[105] O. Teichmüller, *Extremale quasiconforme Abbildungen und quadraticshe Differentiale*, Abh. Preuss. Akad. Wiss., math.–naturw. Kl, **22**, (1939), 1–197, or *Gesammelte Abbildungen–Collected papers*, ed. L.V. Ahlfors and F.W. Gehring, Springer–Verlag, 1982.

[106] W.P. Thurston, *Three-dimensional geometry and topology*. Princeton Mathematical Series, 35, Princeton University Press, Princeton, NJ, 1997.

[107] P. Tukia, *Compactness properties of* μ*–homeomorphisms*, Ann. Acad. Sci. Ser. A I Math., **16**, (1991), 47–69.

[108] I.N. Vekua, *Generalized Analytic Functions*, Oxford, Pergamon Press, 1962.

[109] S. Vodop'yanov, *Topological and geometrical properties of mappings with summable Jacobian in Sobolev spaces*, Sibirsk Mat. Z., **41**, (2000), 23–48.

[110] A. Volberg and F. Nazarov, *Heat extension of the Beurling operator and estimates for its norm*, Algebra i Analiz **15**, (2003), 142–158; translation in St. Petersburg Math. J., **15**, (2004), 563–573.

[111] D.V. Widder, *The Laplace Transform*, Princeton University Press, Princeton NJ, 1946.

Editorial Information

To be published in the *Memoirs*, a paper must be correct, new, nontrivial, and significant. Further, it must be well written and of interest to a substantial number of mathematicians. Piecemeal results, such as an inconclusive step toward an unproved major theorem or a minor variation on a known result, are in general not acceptable for publication.

Papers appearing in *Memoirs* are generally at least 80 and not more than 200 published pages in length. Papers less than 80 or more than 200 published pages require the approval of the Managing Editor of the Transactions/Memoirs Editorial Board.

As of September 30, 2007, the backlog for this journal was approximately 14 volumes. This estimate is the result of dividing the number of manuscripts for this journal in the Providence office that have not yet gone to the printer on the above date by the average number of monographs per volume over the previous twelve months, reduced by the number of volumes published in four months (the time necessary for preparing a volume for the printer). (There are 6 volumes per year, each usually containing at least 4 numbers.)

A Consent to Publish and Copyright Agreement is required before a paper will be published in the *Memoirs*. After a paper is accepted for publication, the Providence office will send a Consent to Publish and Copyright Agreement to all authors of the paper. By submitting a paper to the *Memoirs*, authors certify that the results have not been submitted to nor are they under consideration for publication by another journal, conference proceedings, or similar publication.

Information for Authors

Memoirs are printed from camera copy fully prepared by the author. This means that the finished book will look exactly like the copy submitted.

Initial submission. The AMS uses Centralized Manuscript Processing for initial submissions. Authors should submit a PDF file using the Initial Manuscript Submission form found at www.ams.org/cgi-bin/peertrack/submission.pl, or send one copy of the manuscript to the following address: Centralized Manuscript Processing, MEMOIRS OF THE AMS, 201 Charles Street, Providence, RI 02904-2294 USA. If a paper copy is being forwarded to the AMS, indicate that it is for it Memoirs and include the name of the corresponding author, contact information such as email address or mailing address, and the name of an appropriate Editor to review the paper (see the list of Editors below).

The paper must contain a *descriptive title* and an *abstract* that summarizes the article in language suitable for workers in the general field (algebra, analysis, etc.). The *descriptive title* should be short, but informative; useless or vague phrases such as "some remarks about" or "concerning" should be avoided. The *abstract* should be at least one complete sentence, and at most 300 words. Included with the footnotes to the paper should be the 2000 *Mathematics Subject Classification* representing the primary and secondary subjects of the article. The classifications are accessible from www.ams.org/msc/. The list of classifications is also available in print starting with the 1999 annual index of *Mathematical Reviews*. The Mathematics Subject Classification footnote may be followed by a list of *key words and phrases* describing the subject matter of the article and taken from it. Journal abbreviations used in bibliographies are listed in the latest *Mathematical Reviews* annual index. The series abbreviations are also accessible from www.ams.org/publications/. To help in preparing and verifying references, the AMS offers MR Lookup, a Reference Tool for Linking, at www.ams.org/mrlookup/.

Electronically prepared manuscripts. The AMS encourages electronically prepared manuscripts, with a strong preference for $\mathcal{AMS}$-LaTeX. To this end, the Society has prepared $\mathcal{AMS}$-LaTeX author packages for each AMS publication. Author packages include instructions for preparing electronic manuscripts, samples, and a style file that generates the particular design specifications of that publication series. Though $\mathcal{AMS}$-LaTeX is the highly preferred format of TeX, author packages are also available in $\mathcal{AMS}$-TeX.

Authors may retrieve an author package from the AMS website starting from www.ams.org/tex/ or via FTP to ftp.ams.org (login as anonymous, enter username as password, and type cd pub/author-info). The *AMS Author Handbook* and the *Instruction Manual* are available in PDF format following the author packages link from www.ams.org/tex/. The author package can also be obtained free of charge by sending email to tech-support@ams.org (Internet) or from the Publication Division, American Mathematical Society, 201 Charles St., Providence, RI 02904-2294, USA. When requesting an author package, please specify $\mathcal{AMS}$-LaTeX or $\mathcal{AMS}$-TeX and the publication in which your paper will appear. Please be sure to include your complete mailing address.

After acceptance. The final version of the electronic file should be sent to the Providence office (this includes any TeX source file, any graphics files, and the DVI or PostScript file) immediately after the paper has been accepted for publication.

Before sending the source file, be sure you have proofread your paper carefully. The files you send must be the EXACT files used to generate the proof copy that was accepted for publication. For all publications, authors are required to send a printed copy of their paper, which exactly matches the copy approved for publication, along with any graphics that will appear in the paper.

Accepted electronically prepared files can be submitted via the web at www.ams.org/submit-book-journal/, sent via FTP, or sent on CD-Rom or diskette to the Electronic Prepress Department, American Mathematical Society, 201 Charles Street, Providence, RI 02904-2294 USA. TeX source files, DVI files, and PostScript files can be transferred over the Internet by FTP to the Internet node ftp.ams.org (130.44.1.100). When sending a manuscript electronically via CD-Rom or diskette, please be sure to include a message identifying the paper as a Memoir.

Electronically prepared manuscripts can also be sent via email to pub-submit@ams.org (Internet). In order to send files via email, they must be encoded properly. (DVI files are binary and PostScript files tend to be very large.)

Electronic graphics. Comprehensive instructions on preparing graphics are available at www.ams.org/jourhtml/. A few of the major requirements are given here.

Submit files for graphics as EPS (Encapsulated PostScript) files. This includes graphics originated via a graphics application as well as scanned photographs or other computer-generated images. If this is not possible, TIFF files are acceptable as long as they can be opened in Adobe Photoshop or Illustrator. No matter what method was used to produce the graphic, it is necessary to provide a paper copy to the AMS.

Authors using graphics packages for the creation of electronic art should also avoid the use of any lines thinner than 0.5 points in width. Many graphics packages allow the user to specify a "hairline" for a very thin line. Hairlines often look acceptable when proofed on a typical laser printer. However, when produced on a high-resolution laser imagesetter, hairlines become nearly invisible and will be lost entirely in the final printing process.

Screens should be set to values between 15% and 85%. Screens which fall outside of this range are too light or too dark to print correctly. Variations of screens within a graphic should be no less than 10%.

Inquiries. Any inquiries concerning a paper that has been accepted for publication should be sent to memo-query@ams.org or directly to the Electronic Prepress Department, American Mathematical Society, 201 Charles St., Providence, RI 02904-2294 USA.

Editors

This journal is designed particularly for long research papers, normally at least 80 pages in length, and groups of cognate papers in pure and applied mathematics. Papers intended for publication in the *Memoirs* should be addressed to one of the following editors. The AMS uses Centralized Manuscript Processing for initial submissions to AMS journals. Authors should follow instructions listed on the Initial Submission page found at www.ams.org/memo/memosubmit.html.

Algebra to ALEXANDER KLESHCHEV, Department of Mathematics, University of Oregon, Eugene, OR 97403-1222; email: ams@noether.uoregon.edu

Algebraic geometry and its application to MINA TEICHER, Emmy Noether Research Institute for Mathematics, Bar-Ilan University, Ramat-Gan 52900, Israel; email: teicher@macs.biu.ac.il

Algebraic geometry to DAN ABRAMOVICH, Department of Mathematics, Brown University, Box 1917, Providence, RI 02912; email: amsedit@math.brown.edu

Algebraic number theory to V. KUMAR MURTY, Department of Mathematics, University of Toronto, 100 St. George Street, Toronto, ON M5S 1A1, Canada; email: murty@math.toronto.edu

Algebraic topology to ALEJANDRO ADEM, Department of Mathematics, University of British Columbia, Room 121, 1984 Mathematics Road, Vancouver, British Columbia, Canada V6T 1Z2; email: adem@math.ubc.ca

Combinatorics to JOHN R. STEMBRIDGE, Department of Mathematics, University of Michigan, Ann Arbor, Michigan 48109-1109; email: FRS@umich.edu

Complex analysis and harmonic analysis to ALEXANDER NAGEL, Department of Mathematics, University of Wisconsin, 480 Lincoln Drive, Madison, WI 53706-1313; email: nagel@math.wisc.edu

Differential geometry and global analysis to LISA C. JEFFREY, Department of Mathematics, University of Toronto, 100 St. George St., Toronto, ON Canada M5S 3G3; email: jeffrey@math.toronto.edu

Dynamical systems and ergodic theory to AMIE WILKINSON, Department of Mathematics, Northwestern University, 2033 Sheridan Road, Evanston, IL 60208-2730; email: transactions@math.northwestern.edu

Functional analysis and operator algebras to DIMITRI SHLYAKHTENKO, Department of Mathematics, University of California, Los Angeles, CA 90095; email: shlyakht@math.ucla.edu

Geometric analysis to WILLIAM P. MINICOZZI II, Department of Mathematics, Johns Hopkins University, 3400 N. Charles St., Baltimore, MD 21218; email: trans@math.jhu.edu

Geometric analysis to MLADEN BESTVINA, Department of Mathematics, University of Utah, 155 South 1400 East, JWB 233, Salt Lake City, Utah 84112-0090; email: bestvina@math.utah.edu

Harmonic analysis, representation theory, and Lie theory to ROBERT J. STANTON, Department of Mathematics, The Ohio State University, 231 West 18th Avenue, Columbus, OH 43210-1174; email: stanton@math.ohio-state.edu

Logic to STEFFEN LEMPP, Department of Mathematics, University of Wisconsin, 480 Lincoln Drive, Madison, Wisconsin 53706-1388; email: lempp@math.wisc.edu

Partial differential equations to GUSTAVO PONCE, Department of Mathematics, South Hall, Room 6607, University of California, Santa Barbara, CA 93106; email: ponce@math.ucsb.edu

Partial differential equations and dynamical systems to PETER POLACIK, School of Mathematics, University of Minnesota, Minneapolis, MN 55455; email: polacik@math.umn.edu

Probability and statistics to KRZYSZTOF BURDZY, Department of Mathematics, University of Washington, Box 354350, Seattle, Washington 98195-4350; email: burdzy@math.washington.edu

Real analysis and partial differential equations to DANIEL TATARU, Department of Mathematics, University of California, Berkeley, Berkeley, CA 94720; email: tataru@math.berkeley.edu

All other communications to the editors should be addressed to the Managing Editor, ROBERT GURALNICK, Department of Mathematics, University of Southern California, Los Angeles, CA 90089-1113; email: guralnic@math.usc.edu.

Titles in This Series

895 **Steffen Roch,** Finite sections of band-dominated operators, 2008

894 **Martin Dindoš,** Hardy spaces and potential theory on C^1 domains in Riemannian manifolds, 2008

893 **Tadeusz Iwaniec and Gaven Martin,** The Beltrami Equation, 2008

892 **Jim Agler, John Harland, and Benjamin J. Raphael,** Classical function theory, operator dilation theory, and machine computation on multiply-connected domains, 2008

891 **John H. Hubbard and Peter Papadopol,** Newton's method applied to two quadratic equations in $\mathbb{C}^2$ viewed as a global dynamical system, 2008

890 **Steven Dale Cutkosky,** Toroidalization of dominant morphisms of 3-folds, 2007

889 **Michael Sever,** Distribution solutions of nonlinear systems of conservation laws, 2007

888 **Roger Chalkley,** Basic global relative invariants for nonlinear differential equations, 2007

887 **Charlotte Wahl,** Noncommutative Maslov index and eta-forms, 2007

886 **Robert M. Guralnick and John Shareshian,** Symmetric and alternating groups as monodromy groups of Riemann surfaces I: Generic covers and covers with many branch points, 2007

885 **Jae Choon Cha,** The structure of the rational concordance group of knots, 2007

884 **Dan Haran, Moshe Jarden, and Florian Pop,** Projective group structures as absolute Galois structures with block approximation, 2007

883 **Apostolos Beligiannis and Idun Reiten,** Homological and homotopical aspects of torsion theories, 2007

882 **Lars Inge Hedberg and Yuri Netrusov,** An axiomatic approach to function spaces, spec tral synthesis and Luzin approximation, 2007

881 **Tao Mei,** Operator valued Hardy spaces, 2007

880 **Bruce C. Berndt, Geumlan Choi, Youn-Seo Choi, Heekyoung Hahn, Boon Pin Yeap, Ae Ja Yee, Hamza Yesilyurt, and Jinhee Yi,** Ramanujan's forty identities for Rogers-Ramanujan functions, 2007

879 **O. García-Prada, P. B. Gothen, and V. Muñoz,** Betti numbers of the moduli space of rank 3 parabolic Higgs bundles, 2007

878 **Alessandra Celletti and Luigi Chierchia,** KAM stability and celestial mechanics, 2007

877 **María J. Carro, José A. Raposo, and Javier Soria,** Recent developments in the theory of Lorentz spaces and weighted inequalities, 2007

876 **Gabriel Debs and Jean Saint Raymond,** Borel liftings of Borel sets: Some decidable and undecidable statements, 2007

875 **C. Krattenthaler and T. Rivoal,** Hypergéométrie et fonction zêta de Riemann, 2007

874 **Sonia Natale,** Semisolvability of semisimple Hopf algebras of low dimension, 2007

873 **A. J. Duncan,** Exponential genus problems in one-relator products of groups, 2007

872 **Anthony V. Geramita, Tadahito Harima, Juan C. Migliore, and Yong Su Shin,** The Hilbert function of a level algebra, 2007

871 **Pascal Auscher,** On necessary and sufficient conditions for L^p-estimates of Riesz transforms associated to elliptic operators on $\mathbb{R}^n$ and related estimates, 2007

870 **Takuro Mochizuki,** Asymptotic behaviour of tame harmonic bundles and an application to pure twistor D-modules, Part 2, 2007

869 **Takuro Mochizuki,** Asymptotic behaviour of tame harmonic bundles and an application to pure twistor D-modules, Part 1, 2007

868 **Gelu Popescu,** Entropy and multivariable interpolation, 2006

867 **Vilmos Totik,** Metric properties of harmonic measures, 2006

866 **William Craig,** Semigroups underlying first-order logic, 2006

865 **Nathanial P. Brown,** Invariant means and finite representation theory of $C*$-algebras, 2006

TITLES IN THIS SERIES

- 864 **John M. Lee,** Fredholm operators and Einstein metrics on conformally compact manifolds, 2006
- 863 **M. Lübke and A. Teleman,** The Universal Kobayashi-Hitchin correspondence on Hermitian manifolds, 2006
- 862 **Alberto Canonaco,** The Beilinson complex and canonical rings of irregular surfaces, 2006
- 861 **Leon A. Takhtajan and Lee-Peng Teo,** Weil-Petersson metric on the universal Teichmüller space, 2006
- 860 **Thomas M. Fiore,** Pseudo limits, biadjoints and pseudo algebras: Categorical foundations of conformal field theory, 2006
- 859 **N. Arcozzi, R. Rochberg, and E. Sawyer,** Carleson measures and interpolating sequences for Besov spaces on complex balls, 2006
- 858 **Enrico Valdinoci, Berardino Sciunzi, and Vasile Ovidiu Savin,** Flat level set regularity of p-Laplace phase transitions, 2006
- 857 **Donatella Danielli, Nocola Garofalo, and Duy-Minh Nhieu,** Non-doubling Ahlfors measures, perimeter measures, and the characterization of the trace spaces of Sobolev functions in Carnot-Carathéodory spaces, 2006
- 856 **Vladimir Bolotnikov and Harry Dym,** On boundary interpolation for matrix valued Schur functions, 2006
- 855 **Yevgenia Kashina, Yorck Sommerhäuser, and Yongchang Zhu,** On higher Frobenius-Schur indicators, 2006
- 854 **Noam Greenberg,** The role of true finiteness in the admissible recursively enumerable degrees, 2006
- 853 **Joachim Krieger,** Stability of spherically symmetric wave maps, 2006
- 852 **Viorel Barbu, Irena Lasiecka, and Roberto Triggiani,** Tangential boundary stabilization of Navier-Stokes equations, 2006
- 851 **Jie Wu,** On maps from loop suspensions to loop spaces and the shuffle relations on the Cohen groups, 2006
- 850 **Siegfried Echterhoff, S. Kaliszewski, John Quigg, and Iain Raeburn,** A categorical approach to imprimitivity theorems for C^*-dynamical systems, 2006
- 849 **Katsuhiko Kuribayashi, Mamoru Mimura, and Tetsu Nishimoto,** Twisted tensor products related to the cohomology of the classifying spaces of loop groups, 2006
- 848 **Bob Oliver,** Equivalences of classifying spaces completed at the prime two, 2006
- 847 **Eric T. Sawyer and Richard L. Wheeden,** Hölder continuity of weak solutions to subelliptic equations with rough coefficients, 2006
- 846 **Victor Beresnevich, Detta Dickinson, and Sanju Velani,** Measure theoretic laws for lim-sup sets, 2006
- 845 **Ehud Friedgut, Vojtech Rödl, Andrzej Ruciński, and Prasad V. Tetali,** A Sharp threshold for random graphs with a monochromatic triangle in every edge coloring, 2006
- 844 **Amadeu Delshams, Rafael de la Llave, and Tere M. Seara,** A geometric mechanism for diffusion in Hamiltonian systems overcoming the large gap problem: Heuristics and rigorous verification on a model, 2006
- 843 **Denis V. Osin,** Relatively hyperbolic groups: Intrinsic geometry, algebraic properties, and algorithmic problems, 2006
- 842 **David P. Blecher and Vrej Zarikian,** The calculus of one-sided M-ideals and multipliers in operator spaces, 2006

For a complete list of titles in this series, visit the
AMS Bookstore at **www.ams.org/bookstore/**.